Also by Paul Davies

The Physics of Time Asymmetry
Space and Time in the Modern Universe
The Runaway Universe
The Forces of Nature
Other Worlds
The Search for Gravity Waves
The Edge of Infinity
The Accidental Universe
Quantum Fields in Curved Space (with N. D. Birrell)
God and the New Physics
Superforce
Quantum Mechanics
The Ghost in the Atom (with J. R. Brown)
The Cosmic Blueprint
Fireball
Superstrings: A Theory of Everything? (with J. R. Brown)
The New Physics (editor)
The Matter Myth (with J. Gribbin)
The Mind of God
The Last Three Minutes
Are We Alone?

ABOUT TIME

EINSTEIN'S
UNFINISHED REVOLUTION

Paul Davies

Simon & Schuster

NEW YORK   LONDON   TORONTO   SYDNEY   TOKYO   SINGAPORE

SIMON & SCHUSTER
Rockefeller Center
1230 Avenue of the Americas
New York, NY 10020

SIMON & SCHUSTER and colophon are registered trademarks
of Simon & Schuster Inc.

Designed by Irving Perkins Associates, Inc.

Manufactured in the United States of America

1   3   5   7   9   10   8   6   4   2

Library of Congress Cataloging-in-Publication Data
Davies, P. C. W.
About time : Einstein's unfinished revolution / Paul Davies.
p.   cm.
Includes bibliographical references (p.   ) and index.
1. Space and time.   2. Relativity (Physics)   3. Astrophysics.
4. Einstein, Albert, 1879–1955.   I. Title.
QC173.59.S65D37   1995
530.1′1—dc20   94-40281
CIP
ISBN 0-671-79964-9

Figure 3.1 on page 85 is adapted from a drawing originally published in *Cassier Magazine*, 1906, reprinted in *The Perpetual Motion Mystery* by R. A. Ford (Bradley, Ill.: Lindsay Publications, Inc., 1986), page 42.

I dedicate this book to my long-suffering family.
The time I spent writing it belonged to them.

CONTENTS

PREFACE

This is the second book I have written on the subject of time. The first, published in 1974, was intended for professional physicists. I always intended to write another book on this subject, for the wider public, but somehow I never found the time. At last I have accomplished it.

Fascination with the riddle of time is as old as human thought. The earliest written records betray confusion and anxiety over the nature of time. Much of Greek philosophy was concerned with making sense of the concepts of eternity versus transience. The subject of time is central to all the world's religions, and has for centuries been the source of much doctrinal conflict.

Although time entered science as a measurable quantity with the work of Galileo and Newton, it is only in the present century that it has developed into a subject in its own right. Albert Einstein, more than anyone else, is responsible for this advance. The story of time in the twentieth century is overwhelmingly the story of Einstein's time. Although I have sketched biographical details where appropriate, this book is not a biography of Einstein, because several such have appeared since his centenary in 1979. Nor did I set out to write a systematic and comprehensive study of time. Instead, I have chosen a selection of topics that I personally find particularly intriguing or mysterious, and used them to illustrate the general principles of time as we have come to understand them.

Although Einstein's theory of relativity is nearly a century old, its bizarre predictions are still not widely known. Invariably people learn of them with delight, fear and perplexity. Much of the book is devoted to covering the more straightforward consequences of the theory; the broad conclusion I reach, however, is that we are far from having a good grasp of the concept of time. Einstein's work triggered a revolution in our understanding of the subject, but the consequences have yet to be fully worked out. Much of the theory of relativity remains uncharted territory, and crucial topics, like the possibility of time travel, have only very recently received attention. There are also major problems which hint at deep-seated limitations of the theory; discrepancies concerning the age of the universe and obstacles to unifying Einstein's time with quantum physics are two of the more persistent difficulties. Perhaps more worryingly, Einstein's time is seriously at odds with time as we human beings experience it. All this leads me to believe that we must embrace Einstein's ideas, but move on. The orthodox account of time frequently leaves us stranded, surrounded by a welter of puzzles and paradoxes. In my view, Einstein's time is inadequate to explain fully the physical universe and our perception of it.

The scientific study of time has proved to be disturbing, disorienting and startling. It is also befuddling. I have written this book for the reader with no specialist scientific or mathematical knowledge. Technical jargon is kept to a minimum, and numerical values are avoided except where absolutely necessary. However, I cannot deny that the subject is complex and intellectually challenging. To try and ease the pain a bit, I have resorted to the device of introducing a tame imaginary skeptic, who may from time to time voice the reader's own objections or queries. Nevertheless, you may well be even more confused about time after reading this book than you were before. That's all right; I was more confused myself after writing it.

Many people have helped me shape my ideas of time over the years. I have especially benefited from discussions and debate with John Barrow, George Efstathiou, Murray Gell-Mann, Ian Moss, James Hartle, Stephen Hawking, Don Page, Roger Penrose, Frank Tipler, William Unruh and John Wheeler. Others whose work has influenced me are mentioned in the text. Thanks must also go to my immediate colleagues and friends, who have provided many useful ideas and insights into the subject. These include Diane Addie, David Blair, Bruce Dawson, Roger Clay, Philip Davies, Susan Davies, Michael Draper, Denise Gamble, Murray Hamilton, Angas Hurst, Andrew Matacz, James McCarthy, Jesper Munch, Graham Nerlich, Stephen Poletti, Peter Szekeres, Jason Twamley and David Wiltshire. Last but by no means least is Anne-Marie Grisogono, whose critical reading of the manuscript and challenging discussions of the subject matter have proved invaluable.

Adelaide, South Australia

I saw Eternity the other night,
Like a great ring of pure and endless light,
All calm, as it was bright,
And round beneath it, Time in hours, days, years,
Driv'n by the spheres
Like a vast shadow moved; in which the world
And all her train were hurled. . . .

HENRY VAUGHAN
"The World"

PROLOGUE

Lives of great men all remind us,
We can make our lives sublime,
And, departing, leave behind us
Footprints on the sands of time

H. W. LONGFELLOW

Everybody loves a hero. From Greek mythology to the modern world of pop stars and sporting prodigies, the spectacular achievements of a few individuals have proved far more fascinating than those of the community as a whole. Science is no exception. Aristotle, Galileo Galilei, Isaac Newton, Charles Darwin—these names stand out from the crowd as the shakers and movers of scientific revolutions. Among this roll call of scientific genius, one name best symbolizes both intellectual sparkle and the instigation of dramatic change in our world view: Albert Einstein. Already a legend in his own lifetime, Einstein epitomizes all that the public associates with scientific excellence. He was eccentric-looking and disheveled, he spoke English with a Germanic accent, his theories were cast in arcane mathematics, and he apparently produced his most iconoclastic ideas almost single-handedly, plucking strange new concepts from some abstract Platonic realm and finding that nature obligingly complied.

As with all legends, that of Einstein the scientist contains some measure of truth. He was a genius, he did revolutionize science, and much (though by no means all) of his work was largely a result of his own efforts. He was also stubborn and badly wrong in a number of his scientific ideas. Einstein the man—husband, father, philosopher, musician and world statesman—is a far more complex individual. The idolatry that has surrounded this enigmatic figure for decades is slowly being stripped away a century or so after his birth with the appearance of a number of "warts-and-all" biographies that cast him in a less-than-favorable light as a human being.

Above all, however, Einstein was a man of his time. By the turn of the century, physics had reached a curious crossroads. It was already a mature subject, with well-tried and tested procedures and an impressive track record. In the minds of some enthusiastic physicists, the entire discipline was nearing a state of completion. It was possible to believe that Newton's laws of motion and gravitation, Maxwell's theory of electromagnetism, the laws of thermodynamics and a handful of additional principles might adequately account for all physical phenomena. In that respect, physics at the end of the nineteenth century resembled physics at the end of the twentieth century. An all-encompassing final theory—a Theory of Everything—seemed to be within grasp. Unfortunately, then as now, a few obstinate mysteries darkened an otherwise shining record of success. On the experimental front, the discovery of radioactivity hinted at an energetic world within the atom that lay beyond the scope of gravitation or electromagnetism. The vast age of the Earth, deduced from the fossil record, didn't square with any known process for keeping the sun shining. And the sharp lines in the spectra of gases defied explanation in terms of any realistic model of the atom.

More seriously, inconsistencies in the basic theories themselves lay like hidden reefs waiting to sink the proud ship of "classical" physics. A complete theory of the world cannot be constructed from components that don't mesh together properly. In this respect, two oddities were especially vexing, and eventually forced themselves onto the physicists' agenda.

The first problem concerned the melding of electromagnetic radiation theory and thermodynamics. Each of these subjects was spectacularly successful in its own right. Maxwell's equations of electromagnetism elegantly explained the interweaving of electric and magnetic fields, and provided a theoretical basis for practical devices such as electric motors and dynamos. It also led to the correct prediction of radio waves, and convincingly accounted for the properties of light in terms of electromagnetic waves. The laws of thermodynamics were equally impressive, explaining not only the performance of heat engines, steam engines and refrigerators, but the properties of gases and chemical reactions too. Nevertheless, in the marriage of these two great theoretical schemes there arose a devastating paradox. According to the fashionable picture, space was filled with an invisible substance called the luminiferous ether. Electromagnetic fields were envisaged as strains or distortions in this medium. The trouble was that the hypothetical ether seemed to have an unlimited thermal capacity, an insatiable appetite for heat. Nothing could apparently prevent ordinary matter from progressively surrendering all its heat to the ether in the form of electromagnetic vibrations of arbitrarily high frequency. This apparently inescapable instability implied that material bodies should be unable to retain heat, or to remain in thermal equilibrium with their environment, in blatant contradiction with common sense and experimental evidence.

The second puzzle also had to do with electromagnetism, in this case with the description of moving electric charges. There was a subtle but deep mathematical mismatch between Maxwell's theory of electromagnetism and Newton's laws of motion. Newton's laws were regarded as the founding dictum of physical science, and had long served as a model for all scientific description of change. Formulated in the seventeenth century, they had, by the late nineteenth century, brilliantly withstood the test of time. And yet they came into conflict with electromagnetic theory not just in a technical detail, but in the most basic manner, in the way that they incorporated the concept of motion.

Both these inconsistencies, as I shall explain in the coming chapters, concerned the nature of time. The first—the conflict of electromagnetic theory and thermodynamics—grew out of an attempt to understand the so-called arrow of time, the fact that most physical processes have an inbuilt directionality, manifested especially in the direction of heat flow (from hot to cold). The second involved a clash between Newton's concept of an absolute time and the relativity of motion applied to electrically charged particles.

Before the end of the first decade of the twentieth century, these two theoretical problems had quite simply blown traditional, or classical, physics apart, and set in train not just one but two major scientific revolutions. Stemming from the first puzzle came quantum mechanics, an entirely new and extremely strange theory of matter—so strange that even today many people find it hard to believe: Einstein refused to accept its weird implications throughout his life. The second puzzle gave rise to the theory of relativity. Einstein played a key role in both revolutions, but he is most closely associated with the theory of relativity.

The word "relativity" here refers to the elementary fact that the appearance of the world about us depends on our state of motion: it is "relative." This is obvious in some simple respects even in daily life. If I am standing on the station platform the express train roaring by appears to be moving very fast, whereas if I am riding in the train it is the station that looks to be rushing by. This obvious and uncontentious relativity of motion was known to Galileo and built into Newton's mechanics in the seventeenth century. What Einstein subsequently discovered is that not just motion, but *space and time too* are relative. This was a far more disturbing and counterintuitive claim. As we shall see, Einstein's time challenges our commonsense notions of reality in the most startling manner.

For nineteenth-century scientists it was possible to believe that physics would be complete if it could account for the forces that act between particles of matter, and the way those particles move under the action of the forces. That was all it boiled down to: forces and motion. The particles themselves, and the spacetime arena in which they moved, were just assumed. They were God-given. If nature can be compared to a great cosmic drama in which the contents of the universe—the various atoms of matter—were the cast, and space and time the stage, then scientists considered their job to be restricted entirely to working out the plot.

Today, physicists would not regard the task as complete until they had given a good account of the whole thing: cast, stage and play. They would expect nothing short of a complete explanation for the existence and properties of all the particles of matter that make up the world, the nature of space and time, and the entire repertoire of activity in which these entities can engage. Einstein's greatest contribution was to show that the separation between cast and stage was an artificial one. Space and time are themselves *part of the cast;* they play a full and active role in the great drama of nature. Space and time, as it turns out, are not simply "there" as an unchanging backdrop to nature; they are *physical* things, mutable and malleable, and, no less than matter, subject to physical law.

It took someone of the youth, inexperience, genius and daring of Einstein to question not just the technical correctness but the very conceptual foundation of Newton's physics. Having been successfully battle-tested over two centuries, Newton's concepts of space, time and motion were not to be dismissed lightly. It is a measure of Einstein's greatness that his headlong assault on the edifice of Newtonian physics had become the new orthodoxy within a single generation.

Yet, in spite of devoting his life to the task, Einstein did not succeed in achieving a complete physical theory. He liberated time, and space, from the unnecessarily severe strictures of Newtonian thinking, but was unable to stitch the newly freed concepts of a flexible space and time into a properly unified theory. The search for a unified field theory—or a Theory of Everything, as it is known today—is still at the top of the scientific agenda, and the goal continues to be elusive. Even within the subject of time itself, Einstein left things in a curiously unfinished state. From the dawn of history, the nature of time has proved deeply puzzling and paradoxical to human beings. It is in some ways the most basic aspect of our experience of the world. After all, the very concept of selfhood hinges on the preservation of personal identity through time. When Newton brought time into the domain of scientific inquiry, it proved a fruitful method of analyzing physical processes, but it taught us little about *time itself.*

The scientific, sanitized, picture of time shrugs aside with contempt the accumulated wisdom of traditional cultures in which time is known intuitively, cyclicity and rhythm dominate over measurement, and time and eternity are complementary concepts. The clock, an emblem of our scientific culture, is also the symbol of an intellectual straitjacket. Before Galileo and Newton, time was an organic, subjective thing, not a parameter to be measured with geometrical precision. Time was part and parcel of nature. Newton plucked time right out of nature and gave it an abstract, independent existence, robbing it of its traditional connotations. It was included in Newton's description of the world merely as a way of keeping track of motion mathematically; it didn't actually *do* anything. Einstein restored time to its rightful place at the heart of nature, as an integral part of the physical world. Indeed, Einstein's "spacetime" is in many ways just another field, to be set alongside the electromagnetic and nuclear force fields. It was a monumental first step towards the rediscovery of time.

Important though Einstein's time turned out to be, it still did not solve "the riddle of time." People often ask: What *is* time? Many centuries ago, St. Augustine of Hippo, one of the world's most influential thinkers on the nature of time, gave a perceptive, if enigmatic, reply to this question: "If no one asks me, I know; but if any Person should require me to tell him, I cannot."[1] The time that enters into physical theory, even Einstein's time, bears only the vaguest resemblance to the subjective time of personal experience, the time that we know but cannot explain. For a start, Einstein's time has no arrow: it is blind to the distinction between past and future. Certainly it doesn't *flow* like the time of Shakespeare or James Joyce, or for that matter of Newton. It is easy to conclude that something vital remains missing, some extra quality to time left out of the equations, or that there is more than one *sort* of time. The revolution begun by Einstein remains frustratingly unfinished.

Still, Einstein did make contact with one ancient aspect of time: the traditional association between time and creation. Modern scientific cosmology is the most ambitious enterprise of all to emerge from Einstein's work. When scientists began to explore the implications of Einstein's time for the universe as a whole, they made one of the most important discoveries in the history of human thought: that time, and hence all of physical reality, must have had a definite origin in the past. If time is flexible and mutable, as Einstein demonstrated, then it is possible for time to come into existence—and also to pass away again; there can be a beginning and an end to time. Today the origin of time is called "the big bang." Religious people refer to it as "the creation."

Yet, curiously, Einstein the iconoclast remained so steeped in Newtonian thinking that he did not himself draw this momentous conclusion.

He clung to the belief that the universe is eternal and essentially un-changing in its overall structure, favoring a static cosmology until the accumulating weight of evidence forced him to accept otherwise. But here we encounter the supreme irony. To immobilize his universe, Ein-stein introduced into physics a new type of force, a sort of antigravity. When the universe was shown to be expanding, Einstein discarded the cosmic force with ill-concealed chagrin, later describing it as the biggest blunder of his life. He reluctantly agreed that the universe may not have existed forever, but probably come into being in a big bang some billions of years ago.

Today, the big-bang theory has become the orthodox cosmology. It nevertheless faces a major hurdle in providing a convincing account of how the universe can come to exist from nothing as a result of a physical process. No greater obstacle lies in the path of explanation than the mystery of how *time itself* can originate naturally. Can science *ever* encompass the beginning of time within its scope? This challenge was taken up with panache in the 1980s by a number of theorists, most notably Stephen Hawking, and their efforts were explained to the public in a spate of popular books. Current attempts focus on quantum physics —extended now from a theory of matter to a theory of the whole uni-verse. But time has always lain outside the domain of quantum physics, and attempts to incorporate it end up, paradoxically, by eliminating it. Time vanishes! As I shall explain, there is much about quantum time we do not yet understand.

Despite its popularity, the big-bang theory has not been without its detractors. Right from the start, attempts by astronomers to "date the creation" ran into trouble. The age kept coming out wrong. There wasn't enough time for the stars and planets to come into existence. Worse still, there were astronomical objects that seemed to be older than the uni-verse—an obvious absurdity. Could it be that Einstein's time and cosmic time are not the same? Is Einstein's flexitime simply not flexible enough to stretch all the way back to the creation?

The cosmic-age problem was an embarrassment, and tended to be swept under the carpet, but over the decades it kept popping up again in an irritating manner. In the early years of the subject, cosmologists could wave their arms and make the excuse that their data were still so woolly that a factor of two or three among friends was no reason for a dispute about fundamentals. However, in recent years, with much better tele-scopes and satellite data, cosmology has become almost an exact sci-ence. In 1992, the Cosmic Background Explorer Satellite (COBE) provided what for most cosmologists was the clincher in pinning down the fine details of the big-bang theory. By measuring slight ripples in the background heat of the universe, COBE was able to inject a new level of precision into cosmological modeling. The snag is, the COBE data, combined with other recent observations, have only served to resurrect the age problem with a vengeance.

As I write, the difficulties are hotly debated. Some astronomers believe that, with a bit of fitting and fudging, the time scales can be fixed up. Others disagree and reject the entire big-bang scenario. But a growing number of cosmologists are coming to suspect that the answer may have been provided by Einstein himself. His infamous antigravity force, invented to avoid confronting the origin of time, might just provide the mechanism needed to reconcile it with the extreme ages of certain astronomical objects. His greatest blunder might yet turn out to be his greatest triumph. Time will tell.

CHAPTER 1
A VERY BRIEF HISTORY
OF TIME

Time is at the heart of all that is important to human beings.

BERNARD D'ESPAGNAT

WHOSE TIME IS IT ANYWAY?

Time must never be thought of as pre-existing in any sense; it is a manufactured *quantity.*

HERMANN BONDI

In a dingy laboratory in Bonn lies a submarine-shaped metal cylinder. It is about three meters long, and rests comfortably in a steel frame surrounded by wires, pipes and dials. At first glance, the entire contraption looks like the inside of a giant car engine. In fact, it is a clock—or, rather, *the* clock. The Bonn device, and a network of similar instruments across the world, together constitute "the standard clock." The individual instruments, of which the German model is currently the most accurate, are cesium-beam atomic clocks. They are continually monitored, compared, tweaked and refined via radio signals from satellites and television stations, to cajole them into near-perfect step. At the International Bureau of Weights and Measures at Sèvres, not far from Paris, the data are collected, analyzed and broadcast to a time-obsessed world. Thus originates the famous pips, the radio time signals by which we set our watches.

So, as we go about our daily toil, the Bonn cesium-beam clock keeps the time. It is, so to speak, a custodian of Earth time. The trouble is, the Earth itself doesn't always keep good time. Occasionally our clocks, all supposedly linked to the master system in France like a retinue of obedient slaves, must be adjusted by a second to track changes in the Earth's rotation rate. The last such "leap second" was added on 30 June 1994. The planet's spin, accurate enough to serve as a perfectly suitable clock for a thousand generations, is now defunct as a reliable chronometer. In this age of high-precision timekeeping, poor old Earth doesn't make the grade. Only an atomic clock, man-made and mysterious, serves to deliver those all-important tick-tocks with the precision demanded by navigators, astronomers and airline pilots. One second is no longer defined to be $1/86,400$ of a day: it is 9,192,631,770 beats of a cesium atom.

But whose time is the Bonn clock telling anyway? Your time? My time? God's time? Are the scientists in that cluttered laboratory monitoring the pulse of the universe, fastidiously tracking some abstract cosmic time with atomic fidelity? Might there be another clock, perhaps on another planet somewhere, faithfully ticking out another time altogether, to the joy of its makers?

We know clocks need not agree: the Earth clock gets out of sync with the Bonn clock. So which one is *right?* Well, presumably the Bonn clock, because it's more accurate. But accurate relative to *what?* To *us?* After all, clocks were invented to tell the time for entirely human purposes. Are all humans "on" the same time, however? The patient in the dentist's chair and the audience listening to a Beethoven symphony experience the same atomically tagged duration in quite different ways.

So much of what we believe about time is a result of cultural conditioning. I once met a mystic in Bombay who claimed he could alter his state of consciousness through meditation and so suspend the flow of time altogether; he was unimpressed with talk of atomic clocks. In a lecture in London some years ago, I found myself sharing the platform improbably with the Dalai Lama. Our task was to compare and contrast time as it enters into Western scientific thinking and Eastern philosophy. The Lama spoke with quiet assurance, but unfortunately in Tibetan. Though I tried to follow the translation for enlightenment, I didn't receive much, regrettably. Culture clash, I suppose.

After my lecture, we had a tea break, and the Dalai Lama took my hand as we walked out of the building into the sunshine. Someone dropped to his knees and presented His Holiness with a daffodil, which he graciously accepted. I had the overwhelming impression of a gentle and intelligent man with insights of value to us all, but prevented by the trappings of his office from effectively communicating them to the assembled Western scientists. I came away from the occasion with a deep sense of missed opportunity.

THE QUEST FOR ETERNITY

Eternity! thou pleasing, dreadful thought!

JOSEPH ADDISON

In the madcap world of modern Western society, where time is money, railways, airline schedules, television programs, even cooking are subject to the tyranny of the clock. Our hectic lives are firmly bolted to the treadmill of time. We are slaves of our past and hostages to the future. But was it always thus? Running like a common thread through the history of human thought, East and West, North and South, is a belief that the entire paradigm of human temporality is rooted in some sort of monstrous illusion; it is but an elaborate product of the human mind:

And likewise time cannot itself exist,
But from the flight of things we get a sense of time. . . .
No man, we must confess, feels time itself,
But only knows of time from flight or rest of things.[1]

Thus wrote the Roman poet-philosopher Lucretius in his first-century epic *De Rerum Natura*. From such unsettling ideas it is but a small step to believe that the passage of time can be controlled or even suspended by mental power, as we discover in the following haunting words of the sixteenth-century mystical poet Angelus Silesius:

Time is of your own making,
its clock ticks in your head.
The moment you stop thought
time too stops dead.[2]

For such temporal relativists, true reality is vested in a realm that transcends time: the Land Beyond Time. Europeans call it "eternity," Hindus refer to it as "moksha" and Buddhists as "nirvana." For the Australian aborigines it is the Dream Time. Angelus Silesius again:

Do not compute eternity
as light-year after year
One step across
that line called Time
Eternity is here.[3]

In our struggle to come to terms with mental and physical reality, nothing vexes us more than the nature of time. The paradoxical conjunction of temporality and eternity has troubled Man through the ages. Plato concluded that the fleeting world of daily experience is only half real, an ephemeral reflection of a timeless domain of pure and perfect Forms, which occupy the realm of eternity. Time itself is but an imperfect "moving image of Eternity which remains forever at one," but which we human beings incorrigibly reify: "The past and future are created species of time, which we unconsciously but wrongly transfer to the eternal essence."[4]

The abiding tension between the temporal and the eternal pervades the world's great religions, and has led to generations of heated and sometimes violent theological debate. Is God inside or outside of time? Temporal or eternal? Process or Being? According to Plotinus, a third-century pagan, to exist in time is to exist imperfectly. Pure Being (i.e., God) must therefore be characterized by the utter absence of any relation to time. For Plotinus, time represents a prison for human beings, separating us from the divine realm—the true and absolute reality.

Belief that God lies outside of time altogether also became the established doctrine among many early Christian thinkers, such as Augustine, Boethius and Anselm, starting a tradition that continues to the present day. Like Plato and Plotinus before him, Augustine places God in the realm of eternity, "supreme above time because it is a never-ending present." In this existence, time does not pass; rather, God perceives all times at once:

> Your years are completely present to you all at one because they are at a permanent standstill. They do not move on, forced to give way before the advance of others, because they never pass at all. . . . Your today is eternity.[5]

Thus, the God of classical Christianity not only exists outside of time, but also knows the future as well as the past and present. These far-reaching ideas have been subjected to detailed analysis and received some sharp criticism by the medieval church, as well as by modern theologians and philosophers. The core of the debate is the daunting problem of how to build a bridge between God's presumed eternity on the one hand and the manifest temporality of the physical universe on the other. Can a god who is completely atemporal logically relate in any way at all to a changing world, to human time? Surely it is impossible for God to exist *both* within and outside of time? After centuries of bitter debate, there is still no consensus among theologians about the solution to this profound conundrum. These tangled issues are reviewed in greater depth in my book *The Mind of God,* for those readers who are interested.

ESCAPE FROM TIME

The great thing about time is that it goes on.

ARTHUR EDDINGTON

Although theologians and philosophers wrangle over the technicalities of the logical relationship between time and eternity, many religious people believe that the most powerful insights into the subject are provided, not by academic debate, but by direct revelation:

> I remember that I was going to bathe from a stretch of shingle to which the few people who stayed in the village seldom went. Suddenly the noise of the insects was hushed. Time seemed to stop. A sense of infinite power and peace came upon me. I can best liken the combination of timelessness with amazing fullness of existence to the feeling one gets in watching the rim of a great silent fly-wheel or the unmoving surface of a deep, strongly-flowing river. Nothing happened: yet existence was completely full. All was clear.[6]

This personal story, recounted by the physicist and Anglican bishop Ernest Barnes in his 1929 Gifford Lectures, eloquently captures the combination of timelessness and clarity so often said to be associated with mystical or religious experiences. Can a human being really escape time and glimpse eternity? In Barnes's case, as happens so often in reports from Westerners, the experience came totally out of the blue. But Eastern mystics have perfected special techniques that allegedly can induce such timeless rapture. The Tibetan monk Lama Govinda describes his own experiences thus:

> The temporal sequence is converted into a simultaneous co-existence, the side-by-side existence of things into a state of mutual interpenetration . . . a living continuum in which time and space are integrated.[7]

Many similar descriptions have been published of deep meditation, or even drug-induced mental states, in which human consciousness apparently escapes the confines of time, and reality appears as a timeless continuum.

The Indian philosopher Ruth Reyna believes the Vedic sages "had cosmic insights which modern man lacks. . . . Theirs was the vision not of the present, but of the past, present, future, simultaneity, and No-Time."[8] Sankara, the eighth-century exponent of Advaita Vedanta,

taught that Brahma—the Absolute—is perfect and eternal in the sense
of *absolute timelessness,* and thus the temporal, though real within the
world of human experience, has no ultimate reality. By following the
path of Self-Realization through Advaita, a truly timeless reality may be
attained: "timeless not in the sense of endless duration, but in the sense
of completeness, requiring neither a before nor an after," according to
Reyna. "It is this astounding truth that time evaporates into unreality
and Timelessness may be envisioned as the Real . . . that spells the
uniqueness of Advaita."[9]

The yearning for an escape from time need not involve refined medita-
tive practices. In many cultures it is merely a pervasive yet subconscious
influence—a "terror of history," as anthropologist Mircea Eliade expresses it—which manifests itself as a compulsive search for the Land
Beyond Time. Indeed, this search is *the* founding myth of almost all
human cultures. The deep human need to account for the origin of things
draws us irresistibly back to a time before time, a mythical realm of
timeless temporality, a Garden of Eden, a primordial paradise, its potent
creativity springing from its very temporal contradictions. Whether it is
Athena leaping from the head of Zeus or Mithras slaying the Bull, we
encounter the same heady symbolism of a lost, timeless, perfect realm
that somehow—paradoxically, timelessly—stands in creative relation to
the immediate world of the temporal and the mortal.

This paradoxical conjunction is captured in its most developed form
in the "Dreaming" concept of the Australian aborigines, sometimes referred to as the Eternal Dream Time. According to the anthropologist
W. E. H. Stanner:

> A central meaning of The Dreaming *is* that of a sacred, heroic time long,
> long ago when man and nature came to be as they are; but neither "time"
> nor "history" as we understand them is involved in this meaning. I have
> never been able to discover any aboriginal word for *time* as an abstract
> concept. And the sense of "history" is wholly alien here. We shall not
> understand The Dreaming fully except as a complex of meanings.[10]

Although the Dream Time carries connotations of a heroic past age, it is
wrong to think of that age as now over. "One cannot 'fix' The Dreaming
in time," observes Stanner. "It was, and is, everywhen." Thus the
Dreaming retains a relevance in contemporary aboriginal affairs, because
it is part of the present reality; the "creator beings" are still active today.
What Europeans call "the past" is, for many aboriginal people, *both*
past and present. Stories of creation are often cast in what Europeans
would call the recent past, even as recent as the era of white settlement.
No incongruity is felt, because, for the Australian aborigine, events are
more important than dates. This subtlety is lost on most European

minds; we have become obsessed with rationalizing and measuring time in our everyday lives. Stanner quotes an old Australian black man who expressed this cultural gulf lyrically:

White man got no dreaming.
Him go 'nother way.
White man, him go different,
Him got road belong himself.

The concept of "white man's time" as a "road" down which he marches single-mindedly is an especially apt description, I think, of Western linear time. It is a road that may perhaps lead to progress, but the psychological price we pay for embarking upon it is a heavy one. Fear of death lies at the root of so much we do and think, and with it the desperate desire to optimize the precious duration we have been allotted, to lead life to the full and accomplish something of enduring value. Modern man, wrote J. B. Priestley,

> . . . feels himself fastened to a hawser that is pulling him inexorably toward the silence and darkness of the grave. . . . But no idea of an "eternal dream time," where gods and heroes (from whom he is not separated for ever) have their being, comes shining through to make modern man forget his calendars and clocks, the sands of his time running out.

But even those of us who are trapped within Western culture, for whom a magical, mystical escape route from time is unavailable, can still discern the powerful ancient symbols at work in art and literature, reverberating down the ages. From *Paradise Lost* to *Narnia,* from King Arthur's Avalon to that distant galaxy far away and long ago where the *Star Wars* were fought and won, the realm of eternity has never been very far from the surface. The evocative emblems of eternity now lay shadowy and indistinct in our culture, serving merely as a seductive distraction from the commonsense "reality" of ruthless, passing time. Yet, Priestley assures us, they live on:

> Among the ideas that haunt us—ideas we may laugh at but that will not leave us, ideas that often promise a mysterious happiness when all else seems to fail us—is this one of the Great Time, the mythological dream time, that is behind and above and altogether qualitatively different from ordinary time. We no longer create any grand central system out of it. We do not let it shape and guide our lives. It has dwindled and now looks small and shabby, rather laughable; but it cannot be laughed out of existence, it refuses to go away.[11]

CYCLIC WORLDS AND THE ETERNAL RETURN

All things from eternity are of like forms and come round in a circle.

MARCUS AURELIUS ANTONIUS

In ancient cultures, contact with eternity was kept alive by introducing cyclicity in the world. In his classic text *The Myth of the Eternal Return,* Mircea Eliade describes how traditional societies habitually rebel against the historical notion of time, and yearn instead "for a periodical return to the mythical time of the beginning of things, to the 'Great Time.' "[12] He maintains that the symbols and rituals of ancient cultures represent an attempt to escape from historical, linear, "profane" time, to a mythical or sacred epoch, believing that the suspension of profane time "answers to a profound need on the part of primitive man."[13] Walter Ong, an expert on temporal symbolism, also finds evidence in mythology and folklore for a longing to throw off the trappings of time:

> Time poses many problems for man, not the least of which is that of irresistibility and irreversibility: man in time is moved willy-nilly and cannot recover a moment of the past. He is caught, carried on despite himself, and hence not a little terrified. Resort to mythologies, which associate temporal events with the atemporal, in effect disarms time, affording relief from its threat. This mythological flight from the ravages of time may at a later date be rationalized by various cyclic theories, which have haunted man's philosophizing from antiquity to the present.[14]

Release from historical time may be sought by religious rites, such as the ritual repetition of phrases or gestures that symbolically re-create the original events. Contact with sacred time is often identified with regeneration and renewal. The ancient Festival of New Year, common to both traditional and modern cultures, symbolizes the periodic regeneration or rebirth of nature. In some instances, it represents a repetition of the creation event itself—the mythical transition from chaos to cosmos.

The symbolism underlying these widespread folk practices stems from the ancient belief in temporal cyclicity. Many annual rituals in the Western world have pagan origins that predate Christianity, yet they have been tolerated for centuries by the church. Indeed, cyclic rituals play an important role in the church too, in spite of Christianity's implacable opposition to cyclic time.

Western art, poetry and literature, despite being strongly influenced by the dominance of linear time, nevertheless betrays much hidden and occasionally overt cyclicity. The deep preoccupation with the natural cycle of the seasons, the use of repetitious style, and the liberal employment by writers of a nothing-new-under-the-sun philosophy suggest a fantasized retreat from time's relentless arrow. In some extreme examples, the text itself is structured in a temporally distorting manner, as in James Joyce's *Finnegans Wake,* where the last words of the book run onto the opening passage, or Martin Amis's *The Arrow of Time,* where the entire narrative runs backwards.

Cyclicity retains a deep appeal for some people, yet is abhorrent to others. As we shall see, there is a modern variant of Einstein's cosmology that suggests a cyclic universe, and whenever I give public lectures on cosmology and fail to mention it, somebody inevitably asks me about it. Perhaps the attraction of the model is the prospect of resurrection in subsequent cycles. There is a world of difference, however, between a general sort of cosmic regeneration, and a universe that endlessly repeats itself in precise detail. Plato's assertion of cosmic cyclicity exercised a strong influence on Greek, and later Roman thought. It was taken to the logical extreme by the Stoics, who believed in the concept of *palingenesia*—the literal reappearance of the same people and events in cycle after cycle, an idea that strikes most people today as utterly sterile and repugnant.

NEWTON'S TIME AND THE CLOCKWORK UNIVERSE

I cannot believe that medieval man ever felt trapped in some vast machinery of time.

J. B. PRIESTLEY

The association of time with the mystical, the mental and the organic, fascinating and compelling though it may be, undoubtedly served to hinder a proper scientific study of time for many centuries. Whereas the Greek philosophers developed systematic geometry, and elevated it to a philosophical world view, time remained for them something vague and mysterious, a matter for mythology rather than mathematics. In most ancient cultures, the notion of *timekeeping* cropped up in just a few contexts: in music, in the rhythmic pattern of the seasons and the motions of the heavenly bodies, and in the menstrual cycle. All these topics were overlaid with deep mystical and occult qualities in a way that properties like mass, speed and volume were not.

Aristotle's study of the motion of bodies led him to appreciate the fundamental importance of time, yet he fell short of introducing the notion of time as an abstract mathematical parameter. For Aristotle, time *was* motion. This is hardly revolutionary: we perceive time through motion, whether the movement of the sun across the sky or the hands around a clock face. The concept of time as an independently existing *thing*, an entity in its own right, did not emerge until the European medieval age. The existence of an order in nature has been recognized by all cultures, but it was only with the rise of modern science that a precise and objective meaning could be given to that order. In this quantification, the role of time turned out to be crucial.

On 8 July 1714, the government of Queen Anne determined "That a Reward be settled by Parliament upon such Person or Persons as shall discover a more certain and practicable Method of Ascertaining Longitude than any yet in practice."[15] The prize on offer was the princely sum of £20,000, to be awarded for the construction of a chronometer that was capable of determining longitude at sea to within thirty miles after a six-week voyage. No event better symbolizes the transition from the organic, rhythmic time of traditional folklore to the modern notion of time as a functional parameter with economic and scientific value.

The challenge was taken up by a Yorkshireman named John Harrison, who designed several clocks capable of working at sea. Harrison's fourth instrument, which incorporated a refinement that compensated for temperature changes, was completed in 1759 and submitted for trial two years later. It was conveyed on the ship *Deptford* to Jamaica, where, some two months later, it was found to have accumulated an error of just five seconds. The Admiralty was a bit sticky coming up with the prize money, and by 1765 Harrison had collected only half his reward. He eventually appealed to the King and Parliament, but had turned eighty before he received the balance. Even in the eighteenth century, research funding was tight.

History records that it was Galileo who was foremost in establishing time as a fundamental measurable quantity in the lawlike activity of the cosmos. By measuring the swing of a lamp against the pulse of his wrist while sitting in church, he discovered the basic law of the pendulum— that its period is independent of the amplitude of the swing. Soon the era of precision clockwork was to sweep through Europe, with craftsmen designing ever more accurate timepieces. The push for greater precision in measuring time was not motivated by lofty philosophical or scientific considerations, but by the very practical matter of navigation and trade: sailors need to know the time accurately to be able to compute their longitude from the positions of the stars; the discovery of America, necessitating several weeks of east–west travel, spurred the development of shipborne chronometers.

The crucial position that time occupies in the laws of the universe was not made fully manifest until the work of Newton, in the late seventeenth century. Newton prefaced his presentation with a famous definition of "absolute, true and mathematical time, [which] of itself, and from its own nature, flows equably without relation to anything external." [16] Central to Newton's entire scheme was the hypothesis that material bodies move through space along *predictable* paths, subject to forces which accelerate them, in accordance with strict mathematical laws. Having discovered what these laws were, Newton was able to calculate the motion of the moon and planets, as well as the paths of projectiles and other earthly bodies. This represented a giant advance in human understanding of the physical world, and the beginning of scientific theory as we now understand it.

So successful did Newton's laws of mechanics prove to be that many people assumed they would apply to literally every physical process in the universe. From this belief emerged the picture of the cosmos as a gigantic clockwork mechanism, predictable in its every detail. The clockwork universe enshrined time as a fundamental parameter in the workings of the physical world. This universal, absolute and completely dependable time was the time that entered into the laws of mechanics, and was faithfully kept by the cosmic clockwork. It encapsulated the rule of cause and effect, and epitomized the very rationality of the cosmos. And it gave the world the powerful image of God the Watchmaker.

The great French mathematical physicist Pierre de Laplace, the man who told Napoleon that he "had no need of this hypothesis" when discussing God's action in the Newtonian universe, realized that, if all motion is mathematically determined, then the present state of motion of the universe suffices to fix its future (and past) for all time. In this case, time becomes virtually redundant, for the future is already contained in the present, in the sense that all the information needed to create the future states of the universe resides in the present state. As the Belgian chemist Ilya Prigogine once poetically remarked, God the Watchmaker is reduced to a mere archivist turning the pages of a cosmic history book that is already written. [17] Whereas most ancient cultures viewed the cosmos as a capricious living organism, subject to subtle cycles and rhythms, Newton gave us rigid determinism, a world of inert particles and forces locked in the embrace of infinitely precise lawlike principles.

Newtonian time is in its very essence mathematical. Indeed, starting with the idea of a universal flux of time, Newton developed his "theory of fluxions"—a branch of mathematics better known as the calculus. Our preoccupation with precision timekeeping can be traced to the Newtonian concept of a mathematically precise, continuous flux of time. After Newton, the passage of time became more than merely our stream of consciousness; it began to play a fundamental role in our description of the physical world, something that could be analyzed with unlimited accuracy. Newton did for time what the Greek geometers did for space: idealized it into an exactly measurable dimension. No longer could it be convincingly argued that time is an illusion, a mental construct created by mortal beings from their failure to grasp eternity, because time enters deeply into the very laws of the cosmos, the bedrock of physical reality.

EINSTEIN'S TIME

It was into this world of rigid temporality that Albert Einstein was born. Newton's time had endured for two centuries and was scarcely questioned by Westerners, though it has always rested uneasily alongside Eastern thought, and is alien to the minds of indigenous peoples in America, Africa and Australia. Yet Newton's time is the time of "common sense" (Western style). It is also easy to understand. For Newton, there is but one all-embracing universal time. It is simply *there*. Time cannot be affected by anything; it just goes on flowing at a uniform rate. Any impression of a variation in the rate of time is treated as misperception. Wherever and whenever you are, however you are moving, whatever you are doing, time just marches on reliably at the same pace for everybody, unerringly marking out the successive moments of reality throughout the cosmos.

Among other things, Newton's concept of time invites us to chop it up into past, present and future in an absolute and universal manner. Because the whole universe shares a common time and a common "now," then every observer everywhere, including any little green men on Mars or beyond, would concur with what is deemed to have passed, and what is yet to be. This tidy image of time as defining a succession of universal present moments has important implications for the nature of reality, for in the Newtonian world view only that which happens *now* can be said to be truly real. This is indeed how many nonscientists unquestioningly perceive reality. The future is regarded as "not yet in existence," and perhaps not even decided, while the past has slipped away into a shadowy state of half-reality, possibly remembered but forever lost. "Act, act in the living present!" wrote Longfellow, for it is only the physical state of the world *now* that seems to be concretely real.

But this simple view of time as rigid and absolute—powerful and commonsensical though it may be—is fundamentally flawed. Around the turn of the twentieth century, the Newtonian concept of universal time began yielding absurd or paradoxical conclusions concerning the behavior of light signals and the motion of material bodies. Within a few short years, the Newtonian world view had spectacularly collapsed, taking with it the commonsense notion of time. This profound and far-reaching transformation was primarily due to the work of Einstein.

Einstein's theory of relativity introduced into physics a notion of time that is intrinsically flexible. Although it did not quite restore the ancient mystical ideas of time as essentially personal and subjective, it did tie the experience of time firmly to the individual observer. No longer could one talk of *the* time—only my time and your time, depending on how we are moving. To use the catch phrase: *time is relative*.

Although Einstein's time remained subject to the strictures of physical law and mathematical regulation, the psychological effect of abolishing a universal time was dramatic. In the decades that followed Einstein's original work, scientists probed deeper and deeper into time's mysteries. Might different sorts of clocks measure different sorts of time? Is there a natural clock, or a measure of time, for the universe as a whole? Was there a beginning of time, and will there be an end? What is it that imprints on time a distinct directionality, a lopsidedness between past and future? What is the origin of our sense of the flux of time? Is time travel possible, and if so, how can the paradoxes associated with travel into the past be resolved? Remarkably, in spite of nearly a century of investigation, many of these questions have yet to be satisfactorily answered: the revolution started by Einstein remains unfinished. We still await a complete understanding of the nature of time.

IS THE UNIVERSE DYING?

> *And so some day,*
> *The mighty ramparts of the mighty universe*
> *Ringed round with hostile force,*
> *Will yield and face decay and come crumbling to ruin.*
>
> LUCRETIUS

It is impossible to separate scientific images of time from the cultural background that pervaded Europe during the Renaissance and the modern scientific era. European culture has been strongly influenced by Greek philosophy and the religious systems of Judaism, Islam and Christianity. The Greek legacy was the assumption that the world is ordered and rational, and can be understood through human reasoning: if so, then the nature of time can, in principle, be grasped by mortals. From Judaism came the Western concept of time so central to the scientific world view. In contrast to the pervasive notion of time as cyclic, the Jews came to believe in *linear* time. A central tenet of the Jewish faith, subsequently inherited by both Christianity and Islam, was that of the *historical process*, whereby God's plan for the universe unfolds according to a definite temporal sequence. In this system of belief, the universe was created by God at a definite moment in the past, in a very different state from the one that exists today. The theological succession of events—creation, fall, redemption, judgment, resurrection—is paralleled by a divinely directed sequence of physical events—order out of primeval chaos, origin of the Earth, origin of life, origin of mankind, destruction and decay.

The concept of linear time carries with it the implication of an *arrow* of time, pointing from past to future and indicating the directionality of sequences of events. The origin of time's arrow as a physical principle is still a curiously contentious scientific mystery, to which I shall return in Chapter 9. Scientists and philosophers have been sharply divided over the significance of the arrow of time. The conundrum, put crudely, boils down to this: is the universe getting better or worse? The Bible tells the story of a world that starts in a state of perfection—the Garden of Eden—and degenerates as a result of man's sin. However, a basic component of Judaism, Christianity and Islam is a message of hope, of belief in personal betterment and the eventual salvation of mankind.

In the middle of the nineteenth century, physicists discovered the laws of thermodynamics, and it was soon realized that these implied a universal principle of degeneration. The so-called second law of thermodynamics is often phrased by saying that every closed system tends towards a state of total disorder or chaos. In daily life we encounter the second law in many familiar contexts, well captured by familiar sayings: It's easier to break it than make it; There's no such thing as a free lunch; Sod's Law, Parkinson's Law, etc. When applied to the universe as a whole, the second law implies that the entire cosmos is stuck fast on a one-way slide towards a final condition of *total* degeneration—i.e., maximum disorder—which is identified with the state of thermodynamic equilibrium.

One measure of the remorseless rise of chaos uses a quantity called "entropy," which is defined to be, roughly speaking, the degree of disorder in a system. The second law then states that in a closed system the total entropy can never decrease; at best it remains the same. Almost all natural changes tend to increase the entropy, and we see the second law at work all around us in nature. One of the most conspicuous examples is in the way that the sun slowly burns up its nuclear fuel, spewing heat and light irretrievably into the depths of space, and raising the entropy of the cosmos with each liberated photon. Eventually the sun will run out of fuel and cease to shine. The same slow degeneration afflicts all the stars in the universe. In the mid-nineteenth century, this dismal fate came to be known as the "cosmic heat death." The thermodynamic "running down" of the cosmos represented a significant break with the concept of the Newtonian clockwork universe. Instead of regarding the universe as a perfect machine, physicists now saw it as a gigantic heat engine slowly running out of fuel. Perpetual-motion machines were found to be unrealistic idealizations, and the alarming conclusion was drawn that the universe is slowly dying. Science had discovered pessimistic time, and a new generation of atheistic philosophers, led by Bertrand Russell, wallowed in the depressing inevitability of cosmic doom.

The second law of thermodynamics introduces an arrow of time into the world because the rise of entropy seems to be an irreversible, "downhill" process. By an odd coincidence, just as the bad news about the dying universe was sinking in among physicists, Charles Darwin published his famous book *On the Origin of Species*. Although the theory of evolution shocked people far more than the prediction of a cosmic heat death, the central message of Darwin's book was basically optimistic. Biological evolution also introduces an arrow of time into nature, but it points in the opposite direction to that of the second law of thermodynamics—evolution seems to be an "uphill" process. Life on Earth began in the form of primitive micro-organisms; over time, it has advanced to produce a biosphere of staggering organizational complexity, with millions of intricately structured organisms superbly adapted to their ecological niches. Whereas thermodynamics predicts degeneration and chaos, biological processes tend to be progressive, producing order out of chaos. Here was optimistic time, popping up in science just as pessimistic time was about to sow its seeds of despair.

Darwin himself clearly believed that there is an innate drive in nature towards improvement. "As natural selection works solely by and for the good of each being, all corporeal and mental endowments will tend to progress towards perfection," he wrote.[18] Biologists began to speak about a "ladder of progress," with microbes at the bottom and man at the top. So, although the theory of evolution rejected the idea that God had carefully designed and created each species separately, it left room for a designer God to act in a more subtle way, by directing or guiding the course of evolution over billions of years upwards towards man and maybe beyond.

This progressive philosophy was enthusiastically embraced by several leading European thinkers, such as Henri Bergson, Herbert Spencer, Friedrich Engels, Teilhard de Chardin and Alfred North Whitehead. All saw evidence in the universe as a whole, not just in the Earth's biosphere, of an intrinsic ability for nature to produce order out of chaos. The linear time of these philosophers and scientists was one of faltering, yet ultimately assured, advancement.

Unfortunately, progress in nature did not mesh well with either blind thermodynamic chaos, or the purposeless chaos that supposedly underlies Darwinian evolution. Tension between the concept of a progressive biosphere on the one hand and a universe destined for a heat death on the other produced some confused responses. Some biologists, especially in France, downplayed Darwin's central thesis of random mutations in favor of a mysterious quality called *élan vital*, or life force, responsible for driving organisms in the direction of progress, against the chaotic tendencies of inanimate processes. Belief in such a life force persists in certain nonscientific circles even today. Some philosophers and scientists, worried about the overall fate of the universe, asserted that the second law of thermodynamics could be circumvented under certain circumstances, or should not be applied to the universe as a whole.

The argument still rages. Biologists have long since abandoned the life force, and many argue strenuously that any impression of progress in biological evolution is simply the result of wishful thinking and cultural conditioning. The path of evolutionary change, they claim, is essentially random—"chance caught on the wing," to use Jacques Monod's evocative phrase. Other scientists, many of whom have been influenced by the work of Ilya Prigogine, acknowledge the existence of *self-organizing* processes in nature, and maintain that advancement towards greater organizational complexity is a universal lawlike tendency. Spontaneous self-organization need not conflict with the second law of thermodynamics: such processes always generate entropy as a by-product, so there is a price to be paid to achieve order out of chaos. As far as the ultimate fate of the universe is concerned, which of these counterdirected tendencies—advancing complexity or rising entropy—will win in the end depends crucially on the cosmological model adopted. Those readers with an interest in these eschatological matters may like to read my book *The Last Three Minutes*.

THE RETURN OF THE ETERNAL RETURN

History always repeats itself.

PROVERB

Even as the optimists and pessimists squabbled at the turn of the century about which way the cosmic arrow of time was pointing, the concept of cyclicity made an astonishing entrance into Western science. Physicists were struggling to understand the origin of the laws of thermodynamics in terms of the atomic theory of matter. The most basic thermodynamic process is the flow of heat from hot to cold, a one-way process that epitomizes the second law. In Vienna, Ludwig Boltzmann set out to discover a way of explaining this flow mathematically in terms of molecular motion. He envisaged a vast assemblage of microscopic molecules confined inside a rigid box, rushing about chaotically, colliding with each other and bouncing off the walls of the box.

Boltzmann intended his model to represent a gas. He realized that the random motions of molecules would tend to break up any order, and serve to mix the population of particles very efficiently. For example, the temperature of the gas is determined by the average speed of the molecules, so if at some moment the gas were hotter in a certain region the molecules there would on average move faster than the rest. But this state of affairs would not last for long. Soon the high-speed molecules would collide with the slower-moving particles around them and give up some of their kinetic energy. The excess energy of the molecules from the hot region would diffuse through the entire population until a uniform temperature was reached and the average molecular speed in each region became the same throughout the gas.

Boltzmann backed this plausible physical picture with a detailed calculation in which he applied Newton's laws of motion to the molecules and then used statistical techniques to deduce the collective behavior of large numbers of molecules. He discovered a quantity, defined in terms of the motions of the molecules, that provided a measure of the degree of chaos in the gas. This quantity, Boltzmann proved, always increases in magnitude as a result of the molecular collisions, suggesting it be identified with thermodynamic entropy. If so, Boltzmann's calculation amounted to a derivation of the second law of thermodynamics from Newton's laws.

Shortly after this triumph, a huge hole was knocked in Boltzmann's argument by the French mathematical physicist Henri Poincaré, who rigorously proved that a finite collection of particles confined to a box and subject to Newton's laws of motion must always return to its initial state (or at least very close thereto) after a sufficiently long period of time. The state of the gas therefore undergoes "recurrences." Poincaré's theorem carries the obvious implication that if the entropy of the gas goes up at some stage then it eventually has to come down again so the gas can return to its initial state. Whatever set of molecular motions may increase the entropy, or chaos, of the gas, there must be another set that decreases it. In other words, the behavior of the gas over a long time scale is cyclic. This cyclicity in the state of the gas can be traced to the underlying time symmetry in Newton's laws, which do not distinguish past from future.

The length of Poincaré's cycles are truly enormous—roughly 10^N seconds, where N is the number of molecules (about a trillion trillion in 40 liters of air). The age of the universe is a mere 10^{17} seconds, so the duration of the cycles is huge, even for a handful of molecules. In the case of a macroscopic system, the length of the Poincaré cycles dwarfs all other known time scales. Nevertheless, the cycles are finite in duration, so the possibility of an entropy decrease at some stage in the very far future cannot be denied. Boltzmann's conclusion that entropy can rise only as a result of molecular collisions was therefore shown to be wrong. It was soon to be replaced by a less clear-cut, statistical claim:

that the entropy of the gas will *very probably* rise. Decreases in entropy are possible, as a result of statistical fluctuations. However, the chances of an entropy-decreasing fluctuation fall off very sharply with the size of the fluctuation, implying that large decreases in entropy are exceedingly improbable—but still technically possible. Boltzmann himself went on to suggest that maybe the universe as a whole undergoes Poincaré cycles of immense duration, and that the present relatively ordered state of the universe came about as a result of a fantastically rare decrease in entropy. For almost all the time, the state of the universe would be very close to equilibrium—i.e., the heat-death state. What these ideas suggested is that cosmic heat death was not forever, and resurrection was possible, given long enough.

With the discovery of Poincaré's recurrences, the concept of the eternal return became part of scientific discourse, but in a rather different guise from the folklore version. First, the world takes unimaginably long to return to its present state. Second, the cyclicity involved is not an exact periodicity but merely a statistical recurrence. The situation can be envisaged in terms of card shuffling. If a pack of cards arranged in suit and numerical order is shuffled, then it will almost certainly be in a less ordered state after the shuffling process. However, because the pack has only a finite number of states, continued random shuffling *must* cause any given state to appear and reappear, infinitely often. Simply by chance, the original suit and numerical order will eventually be restored. The state of the cards can be regarded as analogous to the states of the gas, and the shuffling process plays the role of chaotic molecular collisions.

The foregoing argument was seized upon by the German philosopher Friedrich Nietzsche, who concluded that cosmic recurrences robbed human life of any ultimate purpose.[19] The senselessness of endless cycles rendered the universe absurd, he opined. His despairing philosophy of "nihilism" rubbished the concept of progress, whether human or cosmic. Clearly, if the universe is one day to return to its initial state, all progress must eventually be reversed. This conclusion provoked Nietzsche's most famous aphorism: "God is dead!"

THE START OF IT ALL

Einstein was fully aware of the conflicting ideas concerning the arrow of time. Indeed, in the very year he formulated his theory of relativity, he also made a major contribution to the statistical mechanics of molecular motions. Yet, despite this awareness, his first attempt to construct a model of the universe was based on the assumption that it was static and unchanging. In this he was not alone. Most nineteenth-century astronomers believed that the universe remained on average much the same from epoch to epoch. The belief in a stable, eternal cosmos in which degenerative processes are continuously balanced by regeneration dates from the time of ancient Greece. Such models survive to the present day, in the guise of the so-called steady-state theory and its variants.

Cosmologies can thus be divided into four classes. First is the orthodox scientific model of a universe that comes into existence at a finite time in the past and slowly degenerates towards a heat death. Second is a universe that has a definite origin but progresses in spite of the second law of thermodynamics. Third is the cyclic universe with no overall beginning or end, involving either strict repetition or statistical recurrences. Finally there is the static or steady-state universe, in which local processes may be degenerative or progressive but the universe as a whole remains more or less the same forever.

There is no doubt that the widespread acceptance of the first of these cosmological models owes much to Western culture and centuries of entrenched belief in a created universe. This belief brought with it the notion of a universal time—God's time—from which it followed that there must be a definite *date* for the creation. Attempts to deduce the date from an examination of the Bible inevitably gave an answer of a few thousand years B.C. In Renaissance Europe such a figure was not unreasonable. Little was known about geological processes or biological change, still less about the true astronomical arrangement of the universe. It was possible to believe that the universe was just a few millennia old. When the geologists of the nineteenth century pointed to fossils as evidence of Earth's vast age, some churchmen replied that these images were deliberately created by the devil to confuse us. There are religious zealots to this day who declare that we cannot trust our clocks or our senses. They firmly believe the universe was created by God just a few thousand years ago, and merely *looks* old.

Might they be right? Can we be certain the universe really is old? Consider this. The star Sanduleak 69 202 blew up 160,000 years ago, Earth time. Nobody knew this until a technical assistant working at Las Campañas Observatory in Chile saw it happen on the night of 23–24 February 1987. The explosion was clearly visible to the unaided eye in the dark night sky. The news took so long to reach us because Sanduleak 69 202 lies about 1½ billion billion kilometers away, in the nearby minigalaxy known as the Large Magellanic Cloud, and the light from the explosion travels at a finite speed.

If the universe was created a few thousand years ago, it must have been made with Sanduleak 69 202 already in an exploded condition—a star created dead. But that would not be all. In the space between the stricken star and Earth lies a light beam, stretching back from our eyes continuously to the star. And down that beam, marching inexorably towards us, is the record of events which befell the star. Imagine that beam, 160,000 light-years long, on the day of creation. The starbeam, which must have been brought into being intact along with everything else, carries, for the greater part of its length, the image of a dead star, blown to bits, debris flying. But for a short distance near Earth, along a segment just a few thousand light-years long, the beam encodes a curious fiction—images of a living star that never was. The whole contrivance is made simply to *look* as if there was once a living star, whereas in fact God created a dead star.

But how do we know that this bizarre and contrived act of creation happened as long ago as a few thousand years? If God can create a young universe looking old, how can we be sure he did not create it, say, two thousand years ago, perhaps to coincide with the birth of Jesus? This would have meant creating some human records, such as the Old Testament, as well as fossil records such as dinosaurs, and stellar records such as the curiously fixed-up light beam from Sanduleak 69 202. But so what? A Being who can make dead stars can surely fake a few manuscripts. In fact, how can we be sure that the universe wasn't created a hundred years ago, with everything arranged to appear *as if* it were much older? Or, for that matter, perhaps the world started five minutes ago, and we were all made with consistent memories of our earlier activities already in our brains. (Even more interesting would be if our memories varied a bit, to inflame controversies like the number of gunmen who assassinated President Kennedy.)

IT HAPPENS WHEN IT HAPPENS

Time is just one damn thing after another.

ANONYMOUS

When I was a child, I often used to lie awake at night, in fearful anticipation of some unpleasant event the following day, such as a visit to the dentist, and wish I could press some sort of button that would have the effect of instantly transporting me twenty-four hours into the future. The following night, I would wonder whether that magic button was in fact real, and that the trick had indeed worked. After all, it *was* twenty-four hours later, and though I could remember the visit to the dentist, it was, at that time, only a *memory* of an experience, not an experience.

Another button would also send me backwards in time, of course. This button would restore my brain state and memory to what they were at that earlier date. One press, and I could be back at my early childhood, experiencing *once again, for the first time,* my fourth birthday. . . .

With these buttons, gone would be the orderly procession of events that apparently constitutes my life. I could simply jump hither and thither at random, back and forth in time, rapidly moving on from any unpleasant episodes, frequently repeating the good times, always avoiding death, of course, and continuing *ad infinitum*. I would have no subjective impression of randomness, because at each stage the state of my brain would encode a consistent sequence of events.

It is but a small step from this wild fantasy to the suspicion that maybe someone else—a demon or fundamentalist-style deity perhaps—is pressing those buttons in my behalf, and I, poor fool, am totally oblivious to the trickery. On the other hand, so long as the mysterious button-pusher keeps at it, it seems as if I will enjoy some sort of immortality, though one restricted to a fixed set of events. Still, perhaps this is better than mortality? "In eternity there is nothing past and nothing future, but only present," wrote Philo Judaeus.[20] But that was in the first century. We have to be cautious; times have changed since then.

The striking thing about the above "thought experiments" is, how would my life seem any different if this button-pushing business really was going on? What does it even mean to say that I am experiencing my life in a jumpy, random sort of manner? Each instant of my experience *is* that experience, whatever its temporal relation to other experiences. So long as the memories are consistent, what meaning can be attached to the claim that my life *happens* in a jumbled sequence? In his novel *October the First Is Too Late,* the British astronomer and science fiction writer Fred Hoyle also imagined some sort of cosmic button-pusher, but one who fouled things up and got different bits of the world out of temporal kilter. People crossed "time zones" and were bewildered to encounter communities living at different historical periods. Hoyle's fictional scientist caught up in this nightmare has no truck with the notion of time as "an ever-rolling stream," dismissing it as "a grotesque and absurd illusion." He says: "If there's one thing we can be sure enough of in physics it is that all times exist with equal reality."[21] We are invited to think about events in the universe in terms of an unusual metaphor: a series of numbered pigeonholes containing messages about neighboring pigeonholes. The messages accurately describe the contents of the holes with smaller numbers ("the past"), but are vague about those with bigger numbers ("the future"). This mimics causality and the asymmetry between our secure knowledge of the past and woolly predictions of the future. But there is no "flow" of time. Instead there is a metaphorical clerk who inspects the pigeonholes one by one. Each act of inspection creates a moment of consciousness in the world: "As soon as a particular state is chosen, as soon as an imaginary office worker takes a look at the contents of a particular pigeon hole, you have the subjective consciousness of a particular moment, of what you call the present," explains the scientist.

The curious feature of this imagery is that the clerk doesn't need to sample the pigeonholes in numerical sequence. He could capriciously hop about all over the place, even at random, and we wouldn't notice; we would all still have the impression of time as a continuous, ever-rolling stream. Each clerk-activated moment of human consciousness involves a memory-experience of the "pigeonhole contents" further down the numerical sequence, even if the clerk hadn't inspected those pigeonholes for a while. Furthermore, there is nothing to stop the clerk from resampling the same pigeonhole a million times. From the subjective standpoint of the consciousness attached to that pigeonhole, the world appears the same on each go. "It doesn't matter what order you take the pigeonholes," says the scientist, "it doesn't matter if you choose some or all of them a million times, you'd never know anything different from the simple sequential order."

It gets worse. The scientist envisages two rows of pigeonholes. One is for you (i.e., the pigeonholes contain events pertaining to your consciousness), the other me. The clerk gets replaced at this stage in Hoyle's narrative by a less anthropomorphic moving spot of light. "Our consciousness corresponds to just where the light falls, as it dances among the pigeon holes," we are told. But the light does not have to sample (i.e., spotlight) pairs of pigeonholes, one from each row, simultaneously. It could flit back and forth between the rows. There would really be only one consciousness, but two rows of pigeonholes, so the activated consciousness in one row would *feel* different—and regard itself as a different person—from that in the other row. By extension, all conscious beings in the universe, human, animal and alien, could actually be the *same* consciousness, but activated in different contexts at different times. Even if the process was totally random, it would create the impression of an orderly sequence of events being experienced by myriads of distinct minds.

Back in the real world, Pope Gregory XIII pushed a metaphorical button (i.e., issued a decree) in 1582 and the date jumped from 4 October to 15 October overnight. At least it did in Catholic countries. Protestants were suspicious of this Roman sleight-of-hand. Might they be robbed of ten days of their lives? Some confused folk couldn't distinguish *dates* from *times*. Britain and America did not adopt the Gregorian calendar until the eighteenth century; the Russians held out, astonishingly, until 1917. The Pope's adjustment was needed because the Earth does not obligingly circle the sun in an exact number of days; hence the need for leap years. The old Roman calendar did not take accurate enough account of leap years, and the Easter Festival was getting warmer and warmer as the calendar year gradually slid out of synchrony with the seasons. Pope Gregory decreed that century years should not be leap years unless divisible by 400. This rule fixes things up for 3,300 years. More recent refinements to the rule have put us right for another 44,000 years. Rumor has it, though, that the inhabitants of an island in the Outer Hebrides still have no intention of adopting the newfangled Gregorian calendar.

I shall sidestep the issue of psychological pigeonhole time for the moment, and deal with physical, measurable time as if it is real. For that is the founding assumption of science—that there is a real world out there that we can make sense of. And that world includes time. Given a rational universe, we can seek answers to rational questions about time, such as the source of the arrow of time, and the date on which the universe began, if indeed it had a beginning. However, the rational clock-work cosmology of Newton, and the dying thermodynamic cosmology that came after it, were based on a highly simplistic view of time. Though adequate for two hundred years, Newton's conception of time was fundamentally flawed. It took someone of the genius of Albert Einstein to expose its defects.

CHAPTER 2
TIME FOR A CHANGE

From the moment when he came to question the traditional idea of time, only five weeks were needed to write his paper, although he was working all day at the Patent Office.

G. J. WHITROW

A GIFT FROM HEAVEN

Fifteen hundred light-years away, in the constellation of Aquila, lies a bizarre astronomical system. Known cryptically as PSR 1913 + 16, or more simply "the binary pulsar," it consists of a pair of burnt-out, collapsed stars cavorting about each other in a slow-motion dance of death. Each star contains more material than our sun, but squashed into a volume so small it would barely cover Manhattan.

My story of Einstein's time begins with one of those stars. It is spinning several times a second, and as it whirls, so its magnetic field—a trillion times stronger than Earth's—creates a mighty cosmic dynamo. Stray electrons tangle in the magnetic field, and are boosted to almost the speed of light. Yanked protesting into circular paths, the electrons spew forth electromagnetic radiation in a narrow beam. As the star rotates, so the beam sweeps the universe like a lighthouse. Each time it crosses Earth, our radio telescopes can detect a momentary blip. The regular blip, blip, blip of PSR 1913 + 16 marks it out as a very special object—a pulsar. When the first pulsar was discovered, in 1967, it was half-seriously taken to be an alien radio signal, so precise were the pulses. But pulsars are 100 percent natural objects, and scientists soon came to appreciate that their precise radio emissions make them the most accurate clocks in the universe. For example, on 1 September 1974, shortly after its discovery, the pulsation period of PSR 1913 + 16 was determined to be 0.059029995271 second.

In the binary pulsar, the clock star does more than just sit there spinning and bleeping: it also revolves around its companion star in a high-speed orbit. This orbital motion leaves a characteristic imprint in the incessant chatter of radio pulses. The pulsation rate, so regular for a static pulsar, drifts this way, then that, in frequency. Astronomers have eagerly scrutinized every minute detail of this variation, sifting the staccato blips to a fidelity of fifty microseconds. They regard PSR 1913 + 16 as an astronomical gem—so useful and unexpected it has been described as a gift from heaven.

This particular gift was culled from the data of a routine search for new pulsars carried out by a graduate student, Russell Hulse, of the University of Massachusetts at Amherst. Hulse had been sent by his thesis adviser, a young Amherst professor named Joseph Taylor, to spend the summer at Arecibo in Puerto Rico, where the world's largest radio telescope lies sculpted into the landscape. Hulse was lucky to spot the feeble signal on 2 July, because it lay just above the recording threshold. His curiosity aroused by the regular bleeps, Hulse returned in August to observe the object again, and was instantly baffled to find that the period had changed, and kept on changing as he observed. If the object was a pulsar, its pulses were supposed to be absolutely regular. By September, Hulse had discovered that the variations in the period followed a pattern, and he realized the pulsar must be part of a binary star system, the drifting period being caused by the pulsar's orbital motion. It soon became clear that the companion body, like the pulsar, was another collapsed star, and that in PSR 1913 + 16 astronomers had a near-perfect natural laboratory for testing Einstein's theory of relativity. The discovery was deemed worthy enough for Hulse and Taylor to be awarded the 1993 Nobel Prize in Physics.

The closely monitored variations in the pulses from the binary pulsar would be completely incomprehensible were it not for the work of Einstein. An archetypal genius, he played a unique role in the history of science. Indeed, he too might be described as a gift from heaven. Popular images portray him with sloppy clothes, unkempt gray hair and a dreamy gaze. But the Einstein of our present story was a rather dapper and alert twenty-six-year-old, a young man of undoubted promise, but as yet with no truly outstanding accomplishments to his name. And, contrary to the legend, he was no mathematical genius. In fact, Hermann Minkowski, Einstein's mathematics tutor at university, even complained about his poor mastery of mathematics, to the extent of describing him as a "lazy dog." Einstein did, however, possess incisive physical insight.

Born on 14 March 1879 in the German city of Ulm, Einstein was the son of a rather well-bred and artistic mother, Pauline, and a practical-minded businessman father, Hermann. The family was Jewish, but non-practicing, and Albert did not receive a religious upbringing. In fact, at the age of five, he went to a local Catholic school in Munich, where the Einsteins moved in 1880. Albert was not especially happy at his elementary school and did not excel. The headmaster advised Hermann that Albert was unlikely to make a success of anything. Though he was sound and methodical at mathematics, Albert gave no real hint of the formidable scientific prowess to come.

At the age of ten, Albert transferred to Munich's Luitpold Gymnasium. Again, he did not fit in well. The very formal methods of instruction and strong emphasis on classics did not suit his temperament. In fact, he gained more inspiration from his uncle Jakob, his father's partner in a rather shaky electrical-engineering business. Jakob was able to fire the young boy's imagination with conversation and books about science and mathematics. In 1894, when Albert was just fifteen, Hermann and Jakob decided to shift their business to Milan, and left Albert behind to complete his final three years of schooling as a boarder at the Gymnasium. He lasted just six months. Dispirited and unhappy, he resolved to escape. In the event, he was formally expelled from the Gymnasium on the grounds that he was disruptive in class and disrespectful of the teachers.

Turning up unexpectedly in Italy, the rebellious Albert announced to his shocked parents that he intended to renounce both his German citizenship and his Jewish faith. As regards his education, he would attend the highly respected ETH—the Eidgenossische Technische Hochschule—in Zurich. Unfortunately, he failed the entrance examination, and was sent instead to a small cantonal school in the Swiss town of Aarau for a year's coaching. By the end of 1896, Albert had finally managed to gain entry to the ETH, to read science and mathematics. After a largely enjoyable few years as a typical student, bright but inclined to be headstrong, he gained a diploma on 28 July 1900. His average mark was a creditable but not sensational 5 out of 6.

At this stage Einstein became a Swiss citizen, but was passed over for military service on account of his flat feet and varicose veins. Following a brief reunion with his family in Italy, he took a temporary teaching post at a school near Schaffhausen. Meanwhile, he had fallen in love with a young Serbian woman named Mileva Maric, a fellow student at the ETH. By all accounts it was not a suitable match. In July 1901, Mileva announced that she was pregnant, and a family row ensued. She duly gave birth to a daughter, Lieserl, who was quickly sent for adoption and remained a dark secret. Albert and Mileva eventually married, and had two sons.

In the midst of the personal troubles surrounding Mileva's pregnancy, Albert prepared a Ph.D. thesis, and successfully applied for a permanent job at the Swiss Patent Office in Bern. It was from this unlikely location that Einstein began the phase of his working life which, two or three years later, was to shake the foundations of physical science. In 1905— his Annus Mirabilis—Einstein made significant contributions during the space of a few short months to *three* major revolutions in physics. The first of these was the quantum theory, the second statistical mechanics. Einstein's seminal contributions were contained in papers explaining the photoelectric effect and Brownian motion respectively. (Brownian motion is the erratic zigzagging of a tiny particle suspended in a fluid resulting from molecular bombardment. It is named after the biologist Robert Brown, who first observed the effect in pollen grains.)

It is, however, for the third and in many ways the most far-reaching paper that Einstein is mainly remembered. Titled innocuously "On the Electrodynamics of Moving Bodies," it was published in the journal *Annalen der Physik*. The paper consisted of several pages of elementary mathematical reasoning and was aimed at elucidating the behavior of electric charges in motion. Those few short pages were destined to send shock waves through the entire scientific establishment, and set in train a transformation in our understanding of the world that is still incomplete. At the eye of the conceptual storm that Einstein initiated lies the subject of *time*. We shall shortly see that our intuitive, commonsense view of time clashes, hopelessly and blatantly, with the insistent blip, blip, blip of the binary pulsar. Those telltale pulses, as irregular as clockwork, drift and weave their mathematical pattern in a message as clear as the twelve decimal places of precision that attach to their measurement: *Newton's universal time is a fiction.*

The pulsar itself is a dead remnant, the core of a once-bright star that gobbled up its nuclear fuel in a hurry and then, bereft of the vital heat source needed to maintain its internal pressure, collapsed. The core went on collapsing until its density reached a billion metric tons per cubic centimeter. This is the density of matter in the nucleus of an atom, and the pulsar is essentially a gigantic atomic nucleus—a ball of neutrons. In the jargon, it is a "neutron star." Neutron stars are so compact that their gravity is enormous. At the surface, you would weigh billions of times more than on Earth! This is why a neutron star can spin frenetically without flying apart: some are known which rotate over a thousand times a second.

The binary pulsar is unusual because it consists of *two* neutron stars orbiting around each other. Each has a mass of about 1.4 suns. There will be many other such binary systems in the universe, and yet others where a neutron star orbits a black hole. The significance of PSR 1913 + 16 for us is that the pulsar—that clock par excellence—is in an environment where it is subjected to two effects that turn out to be deeply important for our understanding of time: motion and gravitation.

GOODBYE TO THE ETHER

Newton did not suppose that motion could affect time. After all, if time is universal, it cannot depend on whether you, the observer, choose to move or not. In the Newtonian world view, motion (as in the hands of a clock) can be used to *gauge* an all-pervading, already existing time, but not to create it, or to modify it by one iota.

Newton's assumption of temporal solidity was destined to run into grave trouble. Sooner or later, its shortcomings would have manifested themselves to scientists in one way or another. As it happened, events around the turn of the century, both experimental oddities and theoretical paradoxes, brought matters to a head in the field of electromagnetism. It was the motion of electrically charged particles that puzzled Einstein in those early years at the Patent Office. To appreciate what the problem was, you have to understand a concept that was just as important to Newton as his universal time: the *relativity* of motion.

Imagine being inside a box far out in space. You are weightless, and have no sensation of motion. What does it mean to say you are moving? You can look through a window in the box and maybe see a space capsule rush by. Does that mean you are moving, or is it the space capsule that is moving, or both? A radio conversation with the astronaut in the capsule is no help: "I have no sensation of motion either," she says. You are surrounded by space, but there is no way you can tell whether you are moving *through* space, because space contains no landmarks against which you can gauge your motion. It makes clear sense to say you are moving *relative* to the space capsule, but no meaning seems to attach to the notion that you are moving *through space absolutely*.

Newton, and Galileo before him, understood that motion at a uniform speed in a fixed direction is purely relative. On the other hand, *changes* in motion do have absolute effects. If the box you are in suddenly zooms ahead, or veers to one side, you will be flung about and feel forces acting; it will be very noticeable. But no such effects accompany steady, uniform motion in a straight line. You cannot tell, for example, when inside an airliner, whether it is at rest on the ground, or flying at constant velocity in the air. Apart from vibration, it feels exactly the same. Only by looking out the window and seeing whether you are moving relative to the ground can you discern the difference.

Newton incorporated this "principle of relativity" in his laws of motion, and it remained central to physical theory at the turn of the century. For Einstein too, it represented a basic principle of physics, to be retained at all costs. But there was a snag. The laws of electromagnetism, which describe the behavior of electrically charged particles and the motion of electromagnetic waves such as light and radio waves, did not seem to comply with the principle of relativity, yet these laws worked spectacularly well. Crafted by Michael Faraday and James Clerk Maxwell in the mid-nineteenth century, electromagnetic theory had led to the unification of electricity, magnetism and optics, and presaged the modern era of electronics. How could something so *right* be flawed in so basic a fashion?

The clash was at its most glaring with the topic of light propagation. The principle of relativity implied that the speed of light should vary according to the motion of the observer in relation to a light pulse: if you race towards the pulse, it ought to meet you faster than if you try to outrun it. The speed of the light pulse should be meaningful only relative to the reference frame of the observer. On the other hand, electromagnetic theory gave a specific fixed value for the speed of light—about three hundred thousand kilometers per second—with no room for variations depending on motion of the observer. Confusion reigned. Curiously, Einstein had puzzled over this problem even as a teenager, by imagining he could race alongside a light wave. By keeping pace with the wave, he would surely be able to observe the undulating electric and magnetic fields frozen in space around him? Yet that had to be nonsense, because such *static* fields could not exist in empty space without nearby magnets and electric charges to produce them. (*Changing* electric fields can produce magnetic fields, and vice versa.)

The favored resolution of this conflict was to appeal to the ether, a concept I briefly mentioned in the Prologue. This hypothetical medium was supposed to permeate the entire cosmos, filling the space between material bodies. Physicists could then claim it was relative to this ether that light waves traveled at the said constant speed, much as sound waves travel at a certain speed through the air. Now, this ether was evidently peculiar stuff, because it apparently exerted no noticeable mechanical effects—no force or frictional drag—on bodies moving through it. The Earth, for example, could happily plow through the ether on its journey around the sun without feeling the least resistance; and that had better be so, or it would slow up and fall into the sun. Besides being somewhat mysterious, the ether was also an unattractive concept, because it violated the principle of relativity: it implied that a body could be attributed a type of absolute motion in space, even when moving uniformly, if one measured how fast it moved through the ether.

Ugly or not, the idea of an ether was widely accepted. Even today, people sometimes refer to radio signals as "waves in the ether," and spiritualists talk knowingly of "ethereal bodies." But if the ether failed to affect the motion of physical objects, how was its existence to be demonstrated? There is a rule in science that says you should not introduce additional entities unless they exercise some observable physical effect. An invisible substance that never shows up in *any* experiment is an entirely redundant concept. However, in the case of the ether, there did seem to be a way to reveal its ghostly presence. Although it did not affect the motion of the Earth through space, the existence of an ether was relevant to the motion of light. Imagine Earth gliding silently through the invisible ether, at a certain speed in a particular direction. Now suppose there are two oppositely directed light beams, one coming head-on towards Earth through the ether, the other receding from Earth in the same direction as Earth is going. The speed of the former light beam, as measured from Earth, should be greater than that of the latter beam, on account of the Earth's motion. Of course, nobody could say how fast the Earth is moving through space (i.e. through the ether), but we know it circles the sun at about one hundred thousand kilometers per hour, so a speed through the ether of at least this order was feasible.

In the late 1890s, the American physicist Albert Michelson, assisted by Edward Morley, set out to measure the speed of the Earth through space using light beams. To accomplish this feat, they built an apparatus that split a single light beam into two perpendicular beamlets. Each beamlet was directed towards a mirror, and reflected back. The reflected beams were then recombined and examined through a microscope. The theory was this. The Earth is rushing through the ether, so the ether glides past us in a sort of continuous slipstream. We don't feel it, but light does. A light beam that travels against the ether stream would go slower relative to Earth, as I have explained, than one sent downstream. A cross-stream light beam would go at an intermediate speed. Generally, when light pulses are sent out and back in different directions, they should return at slightly different times, because of these differences of speed relative to the laboratory.

Michelson needed to compare the travel times of the beamlets to measure the speed of the ether stream. This is how he did it. Light consists of waves. When the beam is split, the waves of each beamlet start out in step: peak to peak, trough to trough. But when they come back, if the travel times are slightly different, they will be out of step. In the worst case, they will return peak to trough, trough to peak. When the beamlets are recombined, this mismatch shows up: the peaks cancel the troughs, and the troughs cancel the peaks. The effect is to reduce the intensity of the light drastically. This phenomenon is called "destructive interference." So, by monitoring the intensity of light, and swiveling the apparatus around in different directions (the experimenters had no idea which way the ether stream was flowing), Michelson hoped to spot destructive interference and measure the speed of the ether slipstream. This in turn would give a figure for the speed of the Earth through space.

The result of the Michelson-Morley experiment is now a classic in the history of science. The experiment failed to reveal *any evidence at all* for an ether stream. More precisely, the speed of the ether stream was not measurably different from zero. If there is an ether, the Earth is evidently more or less at rest in it. Since this implies that the sun and stars would have to go around the Earth, after the fashion of pre-Copernican cosmology, it was not long before physicists, taking Einstein's lead, decided the ether simply did not exist.

A TIMELY SOLUTION

Was this not revolution indeed? Can one really assert that the time was ripe for a revolution as radical as this? It is probably the greatest mutation ever in the history of human thought.

JEAN ULLMO

How, in the absence of an ether, can the principle of relativity be reconciled with the behavior of light and other electromagnetic phenomena? This was where Einstein made his mark. Before describing his strange and revolutionary solution to the puzzle, let me say a bit about the way in which he thought about physical problems. Einstein was in the deepest sense a theoretical physicist. Though he was, of course, aware of experimental physics, he set much greater store by abstract reasoning. It is not clear that he knew, or cared much, about the now-famous experiment of Michelson and Morley. He merely mentioned in passing in his 1905 paper on the electrodynamics of moving bodies "unsuccessful attempts to discover any motion of the earth relatively to the 'light medium.' "[1]

Einstein has been described as a "top-down" thinker. By this is meant that he began with certain grand, overarching principles which he believed must be true in the real world on account of their philosophical appeal or logical compulsion, and then attempted to project down onto the messy world of observation and experiment to deduce the consequences of these principles. If the consequences appeared strange and counterintuitive at first, then so be it. Mankind has no guarantee from Mother Nature that her secrets will conform to human intuition or notions of common sense. So confident was Einstein in the superiority of human reasoning over empirical observation that when he was once asked what he would have said if his theory had not been confirmed by observation he replied: "I would have had to pity our dear Lord. The theory is correct all the same."[2]

In 1905, Einstein was convinced that the principle of relativity must be upheld at all costs. In this he was greatly influenced by the work of the Austrian philosopher and scientist Ernst Mach, best known for giving his name to the "Mach numbers" for relating speeds to that of sound. Mach belonged to the so-called positivist school of philosophy, which held that reality must be vested in only those things that can be positively observed or detected in some way. For Mach, *all* motion had to be relative (not just uniform motion). The idea that a body such as the Earth could "really" be moving through invisible space was dismissed as meaningless. We tell that a body is moving, claimed Mach, by comparing its position with other bodies, not by imagining it slipping through nothingness.

On the other hand, Einstein did not want to reject the beautiful and successful theory of electrodynamics, with its unique value for the speed of light. So he made a bold leap and retained *both* the relativity of uniform motion *and* the constancy of the speed of light as founding principles of a completely new theory of relativity. Now, these two requirements seem to be plainly contradictory. If motion is relative, then a pulse of light should have a speed that varies relative to the motion of the observer; but then it would not have a constant speed. The only way to effect a reconciliation was to give up something that had been assumed without question since the beginning of science: the universality of space and time. It is easy to see why this step is necessary: it is the only way that two observers who are moving relative to each other can see the *same* pulse of light moving at the *same* speed relative to themselves.

Let me try to illustrate this point in detail. Imagine switching on a flashlight momentarily, and sending a pulse of light off into space. The light will recede from you at 300,000 kilometers per second. Now jump into a rocket ship and zoom after it. Suppose the rocket achieves a speed of 200,000 kilometers per second relative to Earth. Common sense would say that the light pulse is now receding from you at only 100,000 kilometers per second. But, according to Einstein, this is not so: the pulse recedes at 300,000 kilometers per second *both* when you are standing on Earth *and* when you are zooming after the pulse at 200,000 kilometers per second. Whichever reference frame you measure the speed of the pulse from—Earth or rocket—you get the same answer! It doesn't matter how hard you chase the light pulse, you cannot reduce its relative speed by a single kilometer per second. Similarly, if the light pulse is coming towards you, it passes by you at the same speed whether you are at rest on Earth, or zooming towards the light pulse at high speed. A very important corollary of this hypothesis is that the rocket ship could not possibly travel faster than light, for to do so would require the rocket to overtake a receding light pulse, contradicting the assumption that light always recedes from the rocket at the same speed. Because the same principle applies to all observers and reference frames, Einstein's theory implies that no physical object can break the light barrier.

How can we make sense of the apparently absurd state of affairs described above? Speed is distance traveled per unit time, so the speed of light can only be constant in all reference frames if distances and intervals of time are somehow *different* for different observers, depending on their state of motion. The technicalities need not concern us yet. In fact, the mathematics is elementary enough (high-school level), and in his 1905 paper Einstein presented a set of formulae that relate lengths and intervals of time as measured in one frame of reference to their corresponding (different) values when they are observed from another frame of reference. Later I shall give some explicit examples of how these formulae work.

The major upshot of his new theory of relativity, then, was the prediction that time and space are not, as Newton had proclaimed, simply *there,* fixed once and for all in an absolute and universal way for all observers to share. Instead, they are in some sense *malleable,* able to stretch and shrink according to the observer's motion. Einstein came up with this idea of flexitime and elastic space quite suddenly. He had been mulling over the problem of the motion of charged particles for some months, and one day went to see Michele Besso, his good friend from the Patent Office, to use him as a sounding board. Following his extensive discussion with Besso, Einstein found he "could suddenly comprehend the matter." He duly visited Besso the next day and said: "Thank you. I've completely solved the problem." What Einstein had decided was that the commonsense notion of time needed replacing:

> My solution was really for the very concept of time, that is, that time is not absolutely defined but there is an inseparable connection between time and the signal [light] velocity.[3]

Five weeks later, the seminal paper was written and submitted for publication.

Was Einstein right? The binary pulsar was by no means the first opportunity to test Einstein's theory of relativity, but it is definitely one of the best. The pulsar itself moves at about three hundred kilometers per second relative to its companion. The system as a whole moves much more slowly relative to Earth, so the neutron star concerned sometimes rushes towards us, sometimes away. Since the radio pulses it sends out travel at the speed of light (radio and light are both electromagnetic waves and travel at the same speed), here is a system that combines all the salient features of an experiment to test Einstein's theory of relativity: changes of relative motion, light signals, clocks. The signals confirm that even after traveling for fifteen hundred years the pulses coming from the star when it is approaching us have not overtaken those that come from the portion of the orbit when it is receding, proving that the speed of light is independent of the speed of the source. The effects of the distortions of space and time predicted by Einstein's theory are also readily measurable in the precise patterning of the blips. The analysis of the signals is complicated, because they are affected by gravitation as well as motion, but the astronomers have got it all figured out and can untangle the various effects. The bottom line is that Einstein's formulae are validated to very high precision. Time is indeed relative, and can be warped by motion.

Interlude

The morning mail sits opened on my desk. I have half an hour of precious time to peruse it. Among the usual stack of letters, circulars and memoranda are three fat manuscripts sent to me from private addresses: England, California and Western Australia. All came unsolicited and accompanied by letters that start out in the same vein: "Although I am not a scientist . . ." I skim the pages of these manuscripts warily. Like many colleagues, I receive several of them each month. Today they are similar in style and content. Two have some mathematics, handwritten, at pre–high-school level. The message is the same: "Einstein got it wrong; I've got it right. Please help me tell the world."

Closer scrutiny reveals the authors' deep anxieties about time. *How can something so basic to our experience be* relative? *they protest. That would surely lead to paradox. Something must be wrong. The manuscripts contain complicated diagrams showing observers whizzing about with clocks, and agonized questions about whose time is* right *and who is being* misled.

The trouble is, Western culture can't seem to divest itself of the belief in the existence of time as an independently real entity, God-given and absolute. People can accept that clocks may do funny things, and that the human mind may play tricks. But they don't want to attribute such phenomena to time itself, only to the way we experience or measure time. Is this the legacy of "living by the clock," which is so much a feature of modern society? In the old days, men and women were attuned to the cycles and rhythms of nature. They didn't need digital watches to keep appointments by. Train timetables have a lot to answer for: they brought universal precision timekeeping into the lives of ordinary folk. Now a few seconds' error in your watch could cause you to miss the main item on the evening news, or (in Japan at least) the evening train.

Whenever I read dissenting views of time, I cannot help thinking of Herbert Dingle, an irascible but generally well-regarded British philosopher who wrote a book about Einstein's theory of relativity called Relativity for All, *published in 1922. He became Professor of the History and Philosophy of Science at University College London, and must have still been there when I was a student in the UC*

Physics Department between 1964 and 1970. I don't recall ever meeting Professor Dingle, which is probably just as well.

In his latter years, Dingle began seriously to doubt Einstein's concept of time. He had little difficulty persuading a motley group of followers of the absurdity of relative time, and the professor set about attacking the scientific establishment at every opportunity for its adherence to the theory of relativity. Letters were sent to editors in response to routine and innocuous articles about relativity. Editors became exasperated, and rejected the letters. Dingle suspected a conspiracy. He wrote papers for journals pointing out Einstein's errors and had them rejected, too. Dark threats of legal action were rumored. The campaign ended abruptly with Dingle's death, but the mood of dissent he championed lives on, widespread and festering. I wonder why? Einstein must have touched a raw nerve.

STRETCHING TIME

As a theoretical physicist, I rarely find myself performing demonstrations in lectures, but from time to time I take a Geiger counter along. The demonstration is so easy that even I can't make hash of it. I switch the device on, turn up the volume, and wait. Soon the audience gets to hear a random sequence of clicks. That's all. I tell them that the Geiger counter is registering background radiation, most of which is produced by cosmic rays. These are extremely energetic particles from space that bombard the Earth incessantly. Nobody is completely sure what produces them, but if it wasn't for the atmosphere acting as a shield the intensity would be so great the radiation would soon kill us. As it is, the background of cosmic radiation induces mutations in biological organisms, and that helps drive evolution, so in a sense we wouldn't be here without it. But too much would be a bad thing.

Anyway, when these energetic particles strike the nuclei of atoms in the upper atmosphere, they produce all sorts of subatomic debris in great showers. Most of the created particles decay very promptly, but among the longer-lived ones are those that go by the name of "muons." A muon is like an electron, only heavier. Muons don't interact very strongly with ordinary matter, and most of them make it to ground level, some deep into the ground. A lot of the clicks on the Geiger counter are produced by passing muons.

The significance of my little lecture performance is this. If you had a jar of freshly minted muons on your laboratory bench, then after a few millionths of a second nearly all of them would have decayed into electrons. The reason? Muons are inherently unstable, and decay with a half-life of about two microseconds. Now, I have mentioned that no material object can break the light barrier, and this applies to muons as well as anything else: the fastest they could travel is the speed of light. In a few millionths of a second, light travels less than a kilometer, so, on the face of it, the muons created by cosmic-ray impacts at typically twenty kilometers altitude ought not to make it very far towards the ground. Yet the Geiger counter detects them alive and well at ground level.

The explanation lies with time dilation. According to Einstein's theory of relativity, when a muon moves at close to the speed of light, its time becomes highly warped. In our frame of reference fixed to the Earth, moving-muon time becomes considerably stretched out (dilated)—perhaps by a thousand times. Instead of decaying in a few microseconds, Earth-time, a high-speed cosmic-ray muon can live for much longer, long enough to reach the ground. So the clicks in the Geiger counter are auditory testimony to the reality of timewarps.

In checking out the experimental confirmation of the time-dilation effect, I was astonished to discover that the first direct test was not performed until 1941, some thirty-six years after Einstein had first predicted the effect. The experiment, a precision version of the muon business described above, was performed by Bruno Rossi and David Hall of the University of Chicago at two locations near Denver, Colorado. Rossi and Hall wanted to establish that faster muons live longer (as observed by us in the Earth's reference frame). To accomplish this, they deployed metal shields of various stopping powers to filter out the slow muons, and then detected the survivors at two different altitudes, using a bank of connected Geiger counters. They were able to show that the slow particles—which they quaintly referred to as "mesotrons"—decayed about three times quicker than the fast ones. This pioneering work was carried out well after other aspects of special relativity had been thoroughly verified and the theory had long since been accepted by the physics community. In particular, Einstein's famous $E = mc^2$ formula, a well-known by-product of the theory, was by 1941 firmly established; indeed, the atomic-bomb concept, which turns on this formula, was already under investigation in Britain.

Of course, skeptics wouldn't be convinced of time dilation from a few clicks in a Geiger counter. You need to do better than that. Dingle for one wasn't impressed by the experiments. "I do not think that Einstein would have regarded these cosmic ray observations as evidence for his theory," he presumed.[4] In 1972, Dingle published an anguished and caustic attack on belief in the time-dilation effect in particular, and the duplicity of the scientific establishment in general, in a book entitled *Science at the Crossroads,* devoted entirely to rubbishing Einstein's time. "It is impossible to believe that men with the intelligence to achieve the near miracles of modern technology could be so stupid," he fumed.[5] One of the "men" who so irritated Dingle was none other than the eminent Nobel Prize winner Sir Lawrence Bragg, sometime director of the Cavendish Laboratory in Cambridge and of the Royal Institution in London. Bragg was a quiet and methodical Australian physicist, educated at The University of Adelaide, where I now work, who emigrated to England in 1908. Together with his father, William Henry Bragg, Lawrence Bragg developed the important technique of X-ray crystallography, which proved invaluable for unraveling the structure of crystals and, later, organic molecules. Poor Sir Lawrence, in the course of corresponding with Dingle, was rash enough to allude to the fact that cosmic rays appear to observers on the surface of the Earth to last long enough to reach the ground. This excited Dingle's ire, causing him to object strongly to Bragg's "elementary error," pointing out how easy it was to fall into the habit of using words like "mass," "length" and "time" for "hypothetical particles" in the same manner as in daily life. "Physicists have forgotten that their world is metaphorical," he corrected, "and interpret the language literally."[6]

It has to be admitted that many nonscientists share Dingle's skepticism about drawing profound inferences from a chain of mathematical reasoning, in this case concerning "hypothetical" particles that they can't see and which can be detected only with the help of complicated technology. If time really is dilated, they say, let us observe it on *real* clocks. As luck would have it, just a few months before Dingle's polemical tract appeared, a couple of American physicists managed to do just that.

In October 1971, J. C. Hafele, of Washington University in St. Louis, and Richard Keating arranged to borrow four atomic clocks from the U.S. Naval Observatory, where Keating worked. These were cesium-beam clocks made by Hewlett-Packard, of the sort that are used to create our daily time signals. You can't get much closer to Dingle's "everyday" language than that. Hafele and Keating loaded the clocks onto commercial airliners, and gamely flew with them around the world, first towards the east, then towards the west. As airliners travel at less than one-millionth of the speed of light, the timewarp on board was very small indeed—about a microsecond per day's flying. Nevertheless, this level of change was well within the capabilities of the atomic clocks, and the experiment, which must have caused consternation among the other passengers and apoplexy in customs officials, yielded the following results. On the eastward journey, the four clocks came back to America an average of 59 nanoseconds (billionths of a second) slow relative to a set of standard atomic clocks kept at the Observatory. On the westward journey, the clocks were on average 273 nanoseconds fast. The reason for the east–west difference is that, as Einstein noted in his original paper, the rotation of the Earth produces a time dilation too. When the effect of the Earth's rotation was removed, the time dilation produced by the airliners' motion confirmed Einstein's formula.

For those readers who are interested, the said formula is easily given. You take the speed, divide by the speed of light, square the result, subtract it from one and finally take the square root. For example, suppose the speed is 240,000 kilometers per second. Dividing this by the speed of light gives 0.8, squaring gives 0.64, subtracting from 1 gives 0.36 and taking the square root produces the answer of 0.6. So, at a speed of 240,000 kilometers per second, or 80 percent of the speed of light, clocks are slowed by a factor of 0.6, which means they go at 60 percent of their normal rate, or 36 minutes to the hour. I have chosen these numbers because the arithmetic can be done in my head.

The trouble with real clocks, even atomic clocks, is that they are cumbersome and complicated. Once you have decided to accept that the time-dilation effect is real, you can believe that cosmic rays and other high-speed particles also manifest it as claimed. That is to say, Dingle notwithstanding, that our everyday use of words such as "clock" and "time" really can be reliably applied to these indirectly observed sub-atomic entities. It then makes sense to test Einstein's formula using muons rather than atomic clocks, on account of the higher speeds and accuracy that can be attained. In 1966, a group of physicists at the European particle-accelerator laboratory near Geneva, known as CERN, produced some muons artificially and injected them into a ring-shaped vacuum tube where they circulated at 99.7 percent of the speed of light. This had the effect of stretching the muons' time by a factor of twelve or so relative to the lab, so they lived about twelve times as long as they would have done at rest. The controlled nature of the experiment meant that Einstein's time-dilation formula could be checked to within 2 percent accuracy. Of course, it gave the right answer. In 1978, an improved version of the experiment was performed with muons moving still closer to the speed of light, their lifetime being stretched by a factor of twenty-nine.

The experiments leave no shadow of doubt: clocks are affected by motion. But why do physicists insist on concluding that *time* is stretched? The simple answer is: time (for the physicist at least) is that which is measured by clocks. Of course, to be consistent, we must suppose that *all* clocks are affected by motion in exactly the same way; otherwise we would be more inclined to attribute the effect to the clocks rather than to time itself. Well, as far as we can tell, all clocks *are* equally affected (including the brain activity and hence temporal judgment of human observers). This must be so if the principle of relativity is to hold, or we would not have a means of determining whether certain clocks are moving or not—because clocks that are affected differently by motion would get out of synchrony. If you want to drop *that* principle, then all bets are off.

THE PUZZLE OF THE TWINS

So far so good. But now we hit a puzzle. If the motion of clocks is relative, then surely the time-dilation effect is also relative? Suppose we have two clocks, A and B, each in the lap of a human observer, moving relative to each other. In the reference frame of A, it is clock B that is moving, and hence slowed by time dilation. But in the reference frame of B, it is A that is moving, and therefore running slow. So each observer sees the *other* clock running slow! How can that be? It seems like a paradox. If A runs slow, it must fall behind clock B. But if B runs slow, A must *gain* relative to B. How can A be both behind *and* ahead of B at the same time?

This was Dingle's difficulty, in a nutshell. As he wryly remarked, "it requires no super-intelligence to see [it] is impossible."[7] The problem is often referred to as "the twins paradox," because of the following way of expressing it. Imagine a pair of twins, Ann and Betty. Betty goes off in a rocket ship at close to the speed of light and returns to Earth some years later. Ann stays put. Viewed from Earth, Betty's time is slowed, so when Betty gets back Ann should be older than Betty. But viewed from the rocket ship, it is Earth that is moving, so Ann's time is slowed, and on return Betty should find it is *she* who is older. However, both accounts cannot be right: when the twins finally get together again, Betty might either be younger or older than Ann, but not both. Hence the claim of paradox.

In fact, there is no paradox, as Einstein, who first raised the twins problem in passing in his 1905 paper, was quick to realize. The resolution comes from the fact that the two perspectives, Ann's and Betty's, are actually not completely symmetric. To accomplish her trip, Betty must accelerate away from Earth, cruise at uniform speed for a while, then brake, turn around, accelerate again, cruise some more, and finally brake again to land on Earth. Ann merely remains immobile. All Betty's maneuvers, acceleration and deceleration, break the symmetry between the two sets of observations. The principle of relativity, remember, applies to *uniform* motion, not to accelerations. An acceleration is not relative; it is absolute. Taking this into account, it is *Betty* who ages less. On her return, Ann would be older.

It is important to realize two things. First, the twins effect is a *real* effect, not just a thought experiment. Second, it has nothing to do with the effect of motion on the aging process. You must not imagine that the years spent in the rocket ship are somehow kinder to Betty on account of her confinement or movement through space. Suppose for the sake of argument that Betty leaves in the year 2000 and returns in 2020. Ann will have experienced twenty years during Betty's absence, and will of course have aged twenty years as a result. If Betty were to travel at 240,000 kilometers per hour, then, according to Einstein's formula, the journey will take just *twelve* years in her frame of reference. Betty will return, having actually experienced twelve years, and having aged just twelve years, to Earth year 2020. She may be surprised that twenty Earth years have elapsed during *her* twelve years, but her sister's aging will announce it.

The best way to view the twins experiment is in terms of events. There are two delimiting events: Betty's departure from Earth and Betty's return to Earth. Both Ann and Betty must concur on when those events happen, because they witness them together. It is then the case that for Ann twenty years separates the events, whereas for Betty twelve years separates them. There is no inconsistency in this, despite what Dingle may have said. You just have to accept that different observers experience different intervals of time between the same two events. There is no fixed time difference between the events, no "actual" duration, only *relative* time differences. There is Ann's time and Betty's time, and they are not the same. Neither Ann nor Betty is right or wrong in her reckoning; it is just that they differ from each other.

Let me try to give you a better feel for the numbers involved. Suppose you, the reader, are cordially invited to take such a rocket trip, leaving in 2000, returning in 2020. You are given the choice of how quickly you want to "reach" Earth year 2020, which will determine your speed relative to Earth. If you are happy to wait ten years (i.e., to telescope twenty years into ten), you need to travel at 86 percent of the speed of light. To get the duration down to two years, you need to achieve 99.5 percent of the speed of light. I have graphed the relevant relationship in Fig. 2.1. Notice how, the closer to the speed of light you get, the shorter the "trip" between Earth year 2000 and Earth year 2020. The muons in the storage ring at CERN could do it in a few months—if they could live that long.

Hold on, protests the skeptic with a sneaking regard for the Dingles of this world, let's inject some common sense. Suppose you really did this rocket trip; what would you actually *see*? Would the clock on Earth look as if it was running slow or fast, or what? How do the clocks *know* the rocket is going to turn around and come back anyway, thus breaking the symmetry? Whose clock is right?

It is remarkable that, nearly a century after Einstein discovered the relativity of time, people are still thrown by the idea and keep raising the same objections. Even when they get a full explanation, many nonscientists simply don't believe it. So let's take a really good look at a specific

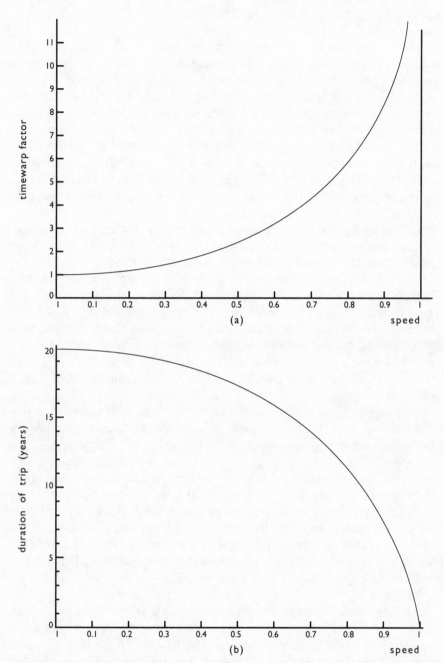

2.1 The time-dilation effect. Graph (a) shows the timewarp factor (i.e., the factor by which a clock is slowed) as a function of speed, expressed as a fraction of the speed of light. For slow speeds the timewarp is small, but as the speed of light is approached it shoots off the top of the graph, becoming infinite at the speed of light itself. Graph (b) shows how the duration of a rocket flight, as experienced in the ship, diminishes with speed. The flight takes twenty years as observed from Earth.

example to try and clear up the matter once and for all. If you don't like technical discussions, I suggest you skip the rest of this section. However, only simple arithmetic plus some imagination is involved.

Betty is going to leave Earth in the year 2000 and travel by rocket ship to a star eight light-years away (as measured in Earth's frame of reference) at a speed of 240,000 kilometers per second. To keep the sums simple, I shall neglect the periods the ship spends accelerating and braking (i.e., treat these periods as instantaneous), and also assume Betty doesn't spend any time sightseeing when she reaches the star. To achieve 80 percent of the speed of light in a negligible time implies an enormous acceleration, which would be fatal to a real human, but this is incidental to the argument. I could easily include a more realistic treatment of the acceleration, but at the price of making the arithmetic more complicated; the overall conclusions would be unaffected.

First let me compute the total duration of the journey as predicted by Einstein for each twin. At 80 percent of the speed of light, it takes ten years to travel eight light-years, so Ann, on Earth, will find that Betty returns in Earth year 2020. Betty, on return, agrees that it is Earth year 2020, but insists that only twelve years have elapsed for her, and her rocket clock—a standard atomic clock carefully synchronized before takeoff with Betty's identical clock on Earth—confirms this assertion: it reads 2012.

Now suppose we equip our twins with powerful telescopes so that they can watch each other's clocks throughout the journey and see for themselves what is going on. Ann's Earth clock ticks steadily on, and Betty looks back at it through her telescope as she speeds away into space. According to Einstein, Betty should see Ann's clock running at 60 percent of the rate of her own clock. In other words, during one hour of rocket time, Betty is supposed to see the Earth clock advance only thirty-six minutes. In fact, she sees it going even slower than this. The reason concerns an extra effect, not directly connected with relativity, that is usually left out of discussions of the twins paradox. It is vital to include the extra effect if you want to make sense of what the twins actually see.

Let me explain what causes this extra slowing. When Betty looks back at Earth, she does not see it as it is at that instant, but as it was when the light left Earth some time before. The time taken for light to travel from Earth to the rocket will steadily increase as the rocket gets farther out in space. Thus Betty will see events on Earth progressively more delayed, because of the need for the light to traverse an ever-widening gap between Earth and rocket. For example, after one hour's flight as measured from Earth, Betty is 0.8 light-hours (48 light-minutes) away, so she sees what was happening on Earth forty-eight minutes earlier, that being the time (as measured in Earth's frame of reference) required for the light, which conveys the images of Earth to Betty, to reach her at that point in the journey. In particular, Ann's clock would appear to Betty—I'm referring to its actual visual appearance—to be slow *anyway*, irrespective of the theory of relativity. After two hours' flight, the Earth clock would appear to Betty to lag even more behind. This "ordinary" slowing down of clocks, and events generally, as seen by a moving observer, is called the "Doppler effect," named for a Swedish physicist who first used it to describe a property of sound waves. By *adding* the Doppler effect to the time-dilation effect, you get the combined slowdown factor.

Ann will also see Betty's rocket clock slowed by the Doppler effect, because light from the rocket takes longer and longer to get back to Earth. She will in addition see Betty's clock slowed by the time-dilation effect. By symmetry, the combined slowdown factor of the other clock should be the same for both of them.

Let me now compute the combined slowdown factor, first from Ann's point of view, then Betty's. To do so, I shall focus on the great event of Betty's arrival at the star. The outward journey takes ten years as measured on Earth. However, Ann will not actually *see* the rocket reach the star in the year 2010, because by this stage Betty is eight light-years away. Since it will take light a further eight years to get back to Earth, it will not be until the year 2018 that Ann gets to witness visually Betty's arrival at the star.

What is the time of the arrival event as registered on Betty's clock? Einstein's formula tells us that Betty's clock runs at 0.6 the rate of the clock on Earth, so ten years of Earth time implies six years in the rocket. The rocket clock therefore stands at six years on Betty's arrival at the star. So, when Ann gets to witness this arrival in 2018, the rocket clock says 2006. Thus, as far as the visual appearance of the rocket clock is concerned, Ann sees only six years having elapsed in her eighteen years—i.e., Betty's rocket clock has been running at *one-third* the rate of Ann's Earth clock. Now, Ann is perfectly capable of untangling the time-dilation and Doppler effects, and computing the "actual" rate of Betty's clock, having factored out the effect of the light delay. She will find the answer to be 0.6, in accordance with Einstein's formula. Thus Ann deduces (but does not actually see) that throughout Betty's outward journey Betty's clock was running at thirty-six minutes to Ann's hour.

From Betty's perspective, things are the other way about. She agrees, of course, that her rocket clock stands at 2006 when she arrives at the star, but what does she see the Earth clock registering at that moment? We know that in the Earth's frame of reference the arrival event occurs at 2010, but, because the star is eight light-years away, the light that actually reaches the rocket at that moment will be from eight years previously—i.e., 2002. So Betty will look back at Earth, on arrival at the star, and see the Earth clock registering 2002. *Her* clock says 2006. Therefore as far as the actual appearance of the Earth clock is concerned, it records two years having elapsed for Betty's six years. Thus Betty concludes that the Earth clock has been running at one-third the rate of her own rocket clock for the outward part of the journey. This is the same factor that Ann perceived Betty's clock to be slowed by, so the situation is indeed perfectly symmetric. Again, Betty can untangle the Doppler effect from the time-dilation effect and deduce that Ann's clock has "really" been running at 0.6 the rate of her own.

Without delay, Betty embarks on the return journey. Because Betty is approaching rather than receding from Earth the light-delay (i.e., Doppler) effect now works in opposition to the time-dilation effect. The former causes events to appear *speeded up*, although time dilation still works to slow them down. Let's put the numbers in. First, what does Ann see as Betty speeds back towards Earth? Since we are agreed that Betty returns to Earth in the year 2020, and Ann actually sees Betty reach the star in 2018, the return part of the journey will appear to Ann, viewing the approach of the rocket from Earth, to be compressed into just two years of Earth time. We have already determined that, when, in 2018, Ann sees Betty's clock at the halfway point, it registers 2006, and that when Betty returns to Earth it will register 2012. So, for the two Earth years during which Ann sees the rocket traveling back, she will witness the rocket clock progress through the remaining six years. In other words, on the return leg of the journey Ann sees Betty's clock running three times *faster* than her own, Earthbound, clock. This is a key point: during the return journey the rocket clock appears from Earth to be *speeded up*, not slowed down. The Doppler effect beats the time-dilation effect. Again, Ann can untangle the time-dilation and light-delay effects and deduce that the rocket clock is "really" running at 0.6 the rate of her clock—i.e., although the rocket clock *looks* to Ann to be speeded up, she *deduces* that it is "really" running slow at exactly the same reduced rate (0.6) as it was on the outward journey. So, although the visual appearance of the rocket clock is quite different for the two legs of the journey, the time-dilation factor of 0.6 remains the same throughout.

Finally, let me examine the return journey as observed by Betty, in the rocket. She has experienced six years for the outward trip, and she experiences another six years for the return, reaching Earth in 2012 as registered on her own clock. During the return journey, however, Betty also observes the clock on Earth. She saw it (actually, visually) standing at 2002 at the moment she reached the star. We know she will get home in 2020, so Betty will see the Earth clock progress through eighteen years during the six years aboard the rocket. Thus the Earth clock appears to Betty to be running three times *faster* than her own rocket clock. This is the *same* factor as that by which Ann saw Betty's clock speeded up—there is complete symmetry on the return part of the journey too. Betty can again factor out the light-delay effect and deduce that the Earth clock is "really" running *slow*—at 0.6 of the rate of her rocket clock.

The crucial point to be extracted from all this is that during the periods when the rocket is traveling at a fixed speed Ann *deduces* that Betty's clock is running slow and Betty *deduces* that Ann's clock is running slow. On the outward part of the journey, each actually *sees* the other's clock running (even more) slowly, but on the return part of the journey each *sees* the other's clock speeded up. The deductions and experiences all fit together consistently, and refute the claim that there is any paradox attached to the statement that "each clock runs slow relative to the other."

For those readers who have waded through this arithmetic, it contains a hidden conclusion about distances. If you use the fact that in Betty's frame of reference Earth recedes at 0.8 of the speed of light, and the journey to the star takes just six rocket years, then the *distance* to the star as measured by Betty must be $0.8 \times 6 = 4.8$ light-years. Thus, although Ann measures the star to be eight light-years away, Betty measures the distance to the star to be only 4.8 light-years. The distance is shrunk by the same factor (0.6) as that by which time is dilated.

GOODBYE TO THE PRESENT

The present moment is a powerful goddess.

JOHANN GOETHE

Although at the end of the journey both Ann's and Betty's experiences dovetail consistently, you can still get in a tangle by asking questions such as: What is Betty doing when Ann's clock registers 2007? Or: What time is it on Ann's clock when Betty arrives at the star? When events occur at spatially separated locations and involve observers in different states of motion, no unambiguous meaning can be attached to these questions. To make them precise, you have to specify exactly which observer, and what sort of observation, you are referring to. When clocks get out of step, there is no universal "now" or present moment on which the different observers can agree. Ann has her definition of "now" in, say, 2007, and Betty has hers. They generally do not match up. For example, you can't expect consistent answers to speculative musings like the following:

Ann: "It's 2007 here on Earth. I wonder if Betty has arrived at her star yet. I know it will only take her six years of her time and she's been gone seven years of my time. Of course, if I look through the telescope, I will see the rocket still well short of the destination, but I know that the telescope doesn't keep me up to date, because of the fact that light from the rocket takes a while to reach me here on Earth. What I want to know is where Betty is *now.*"

In Ann's frame of reference, Betty is $7 \times 0.8 = 5.6$ light-years away, having breakfast at that moment (Ann's 2007 "now"), but of course for Betty that particular breakfast does not occur in 2007. Her own clock reads $7 \times 0.6 = 4.2$ years after departure. If she looks back at Earth, she will actually see the Earth clock reading $4.2 \times \frac{1}{3} = 1.4$ years, but of course she knows it isn't "really that time" on Earth at that moment. To compute *that* date, she has to add the time delay, which is 5.6 years as measured by Ann on Earth. So Betty computes $1.4 + 5.6 = 7$, correctly deducing the date of 2007 on Earth, which *Ann* regards as simultaneous with that particular breakfast in the rocket ship. However, Betty herself sees things differently. She's been gone only 4.2 years in her frame of reference, so the light can't have taken 5.6 of *her* years to reach her from Earth—she hadn't even left then. Since Betty sees Earth receding at 80 percent of the speed of light, in 4.2 years it will be 3.36 light-years away in her frame. It takes 3.36 of her years for the light to reach the rocket from Earth. But because Betty sees the Earth's clock running at $\frac{1}{3}$ the speed of her own, she judges only $3.36/3 = 1.12$ years to have elapsed on Earth since the light was emitted at 1.4 years. This means, as far as Betty is concerned, that the date on Earth "now" (i.e., as she contemplates this vexing matter over that particular meal) is $1.4 + 1.12 = 2.52$ years after departure—definitely *not* 2007. The same figure can be computed without worrying about the light signals by simply noting that the time that has elapsed on Earth since Betty's departure is 0.6 of her own—i.e., $4.2 \times 0.6 = 2.52$. The same arithmetic applied to the date of Betty's arrival at the star (after six rocket years) tells us that this event is simultaneous with 2003.6 years on Earth. By contrast, Ann regards that same event as simultaneous with 2010. The conclusion of all this is that Ann and Betty do not share common "nows." A "Betty-event" B may be regarded *by Ann* as simultaneous with an "Ann-event" A, even though Betty does *not* regard A and B as simultaneous, but picks some entirely different (in this case earlier) Ann-event for that role.

But that's stupid, interjects our skeptic. What happens if Ann calls up Betty and simply *asks* her what she is doing "now"?

She can't do that! The same theory of relativity that predicts the twins effect also forbids any physical body or influence to travel faster than light, so there can be *no* instantaneous communication between Ann and Betty. The fact that Ann and Betty have inconsistent "nows" or definitions of simultaneity at distant places is therefore not a cause for concern. No physically significant meaning can be attached to events happening "now" at a far-flung place, because we can never know about or affect such events in any way. Computing distant "now-events" is purely a bookkeeping exercise. As soon as Ann and Betty are reunited, they can compare notes, and as we have seen they then find their respective stories to be perfectly consistent. If the nonexistence of a universal, agreed, overall "now" strikes you as a crazy idea, it's not new. In 1817, the English essayist Charles Lamb wrote, with uncanny prescience: "Your 'now' is not my 'now'; and again, your 'then' is not my 'then'; but my 'now' may be your 'then', and vice versa."[8]

I have belabored the Ann-Betty saga because I am continually receiving letters asking for clarification of the twins effect, or manuscripts claiming it is false because there is an inconsistency. For those readers who have had the stamina to work through my figures, I hope you will agree that everything hangs together perfectly. There is no paradox. I sincerely hope this is the last word that needs to be said on the subject, though no doubt a few diehard antirelativists will feel moved to pen their objections to my sums.

TIME IS MONEY

How can we be *sure* Einstein is right about the time-dilation effect? In my view, the acid test of any weird theory is this: can you make money from it? One reason I have always been skeptical about the so-called paranormal is that it seems to me, if some people can, say, foretell the future, then they can outperform the average fund-manager on the stock market. Even if the effect is very weak, the gains should still beat the losses over time. Somebody would be marketing the technique by now, and getting very rich. Darwin taught us how even a very slight advantage can exponentiate over time into fantastic success. Sadly, there is slender evidence for paranormal financial acumen among professed psychics (apart from their knack of separating clients from their money, that is). In fact, I recently learned of a clairvoyant who regularly advises top businessmen and politicians, yet who managed to lose the family fortune in the local casino. I do, however, keep an open mind about dowsing, because dowsers can successfully make a living from finding water, in their slightly hokey way.

By contrast with clairvoyance, people regularly make money from dilating time. Engineering handy timewarps has become a commercial operation in several countries. The machine that does it for you is called a "synchrotron." It works by whirling electrons around an evacuated circular tube at very close to the speed of light. Because the electrons are forced along a curved path, they emit intense electromagnetic radiation, concentrated into a narrow beam. (Incidentally, it is this "synchrotron radiation" that explains the pulsar bleeps.) When it was first encountered, synchrotron radiation was a nuisance. Synchrotrons were originally designed to accelerate subatomic particles, not to make radiation. Radiation costs energy, hence money. One reason that particle accelerators are so big is to reduce the curvature of the particles' paths to minimize radiation losses. But, as so often in science, a sin can be turned into a virtue, and today a number of governments have built synchrotrons deliberately to produce the radiation. Synchrotron radiation is very intense, continuously spans a wide range of frequencies from visible light upwards, and is easily manipulated.

The big advantage stems from the very high frequencies that can be achieved—well into the X-ray region of the spectrum. Synchrotron X-rays are used to great effect to elucidate the atomic structure of complicated materials, such as glasses, or large biological molecules. The images are produced so quickly that scientists can sometimes follow the details of chemical changes over time. Recently a group from Wellcome Biotech and Oxford University worked out the structure of the virus that causes foot-and-mouth disease in cattle using the British synchrotron facility at Daresbury in Cheshire. Successful analysis has also been achieved in the field of drug design, thermoplastics and ceramics, and synchrotron lithography has been used to make micromachines less than a millimeter in size. Companies are prepared to pay several thousand dollars a day to use a synchrotron, and these machines turn over millions of dollars a year in commercial activity.

Synchrotron electrons typically travel at 99.99999 percent of the speed of light, and the secret behind their success lies with the time-dilation factor, which is as high as several thousand. This enormously boosts the frequency of the radiation as observed in the reference frame of the laboratory. At low speeds, when relativistic effects can be neglected, electrons in a synchrotron emit radiation at a frequency equal to their circulation frequency around the machine. At high speeds, however, the time-dilation and related relativistic effects make a dramatic difference. The Daresbury synchrotron is ninety-six meters in circumference, and the electrons complete three circuits every microsecond in the laboratory frame: this is in the megahertz frequency range, corresponding to the radio-wave region of the electromagnetic spectrum. A source at this frequency would be useless for studying the atomic structure of materials. But in the reference frame of the electrons, the journey is completed very much faster, because of time dilation, and it is radiation at this higher frequency which is emitted. All told, the effects of relativity increase the frequency of the radiation as observed in the lab to as much as a trillion megahertz.

You don't have to travel to your nearest synchrotron to witness time dilation at work. Its weird effects are subtly manifested in the everyday world about us, because we are surrounded by objects that are moving at extremely high speeds. These objects are electrons circulating within atoms. A typical electron orbits a hydrogen atom at about 200 kilometers per second, or less than one percent of the speed of light. However, the speed is much higher for heavier atoms, on account of the greater electric charge on the nucleus. The inner electrons within atoms such as gold, lead or uranium can whirl around the nucleus at an appreciable fraction of the speed of light. Consequently, the influence of time dilation and other relativity effects will modify the behavior of these electrons in important ways.

To obtain a complete understanding of the electrical and optical properties of solid materials such as gold, physicists must take into account the time dilation of the atomic electrons, including those near the nucleus. For example, take the color of gold. Most metals have a silvery appearance, but not gold. Its distinctive and attractive glitter can be traced to the effects of relativity on the motions of the electrons inside the metal that are responsible for reflecting light. So it is no exaggeration to say that this precious metal is precious—and financially valuable— partly as a result of time dilation operating within the gold atoms.

Many areas of high technology also either depend on, or are affected by, time dilation in one way or another. Radar transmitters, satellite navigation systems and solid state devices are all sensitive to the effects of relativity. Even the humble pips that announce the time signal, produced as they are by finely tuned atomic clocks, would be significantly adrift if the time dilation effect within the clocks themselves were overlooked. So it is that in a host of very practical and even commercial ways, the stretching of time and related aspects of the theory of relativity intrude into our lives.

If time dilation is a real, moneymaking phenomenon, then I am forced to accept (concedes the skeptic) that Ann's now and Betty's now can get out of step. That means my now and your now can also get out of step. But if there's more than one now, isn't there more than one reality? What happens, then, to the orderliness of the universe?

Good question! What sense can we make of physical reality when there is a multiplicity of nows?

TIMESCAPE

The distinction between past, present and future is only an illusion, even if a stubborn one.

ALBERT EINSTEIN

Most Westerners grow up with the firm conviction that reality is vested in the events of the present moment. The basic division of time into past, present and future seems as fundamental to our experience of reality as anything. The past, although remembered, we deem to have slipped out of existence, whereas the future, unknown and mysterious, has yet to be conjured into being. It is a view of the world well captured by the German philosopher Arthur Schopenhauer, who wrote: "The most insignificant present has over the most insignificant past the advantage of reality."[9] Such a belief is not to be dismissed lightly. After long and deep deliberation, as great a thinker on temporal matters as Augustine arrived at precisely this "commonsense" position:

> How can the past and future be when the past no longer is and the future is not yet? As for the present, if it were always present and never moved on to become the past, it would not be time but eternity.[10]

The trouble about common sense is that it can often let you down. After all, common sense suggests that the sun and stars revolve around the Earth. Einstein once remarked that "common sense is that layer of prejudices laid down in the mind prior to the age of eighteen."[11]

The theory of relativity does not imply that you can use a rocket trip to jump into your *own* future, only into somebody else's. You cannot, by changing your state of motion, alter your *here*-and-now, only your *there*-and-now. The mismatch between the "nows" of Ann and Betty refers to what each deduces the *other* is doing "at that moment" at what is in each case a distant place. When the twins get together, their nows coincide once more.

You don't need a rocket ship to dislocate your there-and-now rather violently, if the "there" is far enough away, because the effect increases with distance. Suppose you put down this book, get up from your chair, and walk across the room. You have just changed your there-and-now in the Andromeda galaxy by a whole day! What I mean by this statement is that, while sedentary, you can deduce that a certain event E on a certain planet in Andromeda is happening at the same moment (as judged by you, in your particular frame of reference) as the act of "you reading this passage." When you walk across the room, the event on that distant planet which is simultaneous with your ambulation suddenly changes from "just after E" to some other event which differs from E by a day. It jumps either into the future or past of E, depending on whether you are walking towards or away from Andromeda at the time. Thus simultaneity, like motion, is relative.

So the *time order* of two events can be changed at whim? Doesn't that mean we have the power to reverse time simply by ambling about?

Yes and no. If two events occur at different places (e.g., one on Earth, another in Andromeda), then the time sequence of the two events *can* be reversed, but only if the two spatially separated events occur close enough in time so that light can't get from one to the other in the duration available. Consequently there can be no causal connection between the events, because, according to Einstein, no information or physical influence can travel faster than light between the events to link them causally. So reversing the time order in this restricted case isn't serious: it can't reverse cause and effect, producing causal paradoxes, because the events concerned are completely causally independent. However, this limited ambiguity in the time order of spatially separated events does have an important implication. If reality really is vested in the present, then you have the power to change that reality across the universe, back and forth in time, by simple perambulation. But, then, so does an Andromedan sentient green blob. If the blob oozes to the left and then the right, the present moment on Earth (as judged by the blob, in its frame of reference) will lurch through huge changes back and forth in time.

Unless you are a solipsist, there is only one rational conclusion to draw from the relative nature of simultaneity: events in the past and future have to be every bit as real as events in the present. In fact, the very division of time into past, present and future seems to be physically meaningless. To accommodate everybody's nows—Ann's, Betty's, the green blob's, yours and mine—events and moments have to exist "all at once" across a span of time. We agree that you can't actually witness those differing there-and-now events "as they happen," because instantaneous communication is impossible. Instead, you have to wait for light to convey them to you at its lumbering three hundred thousand kilometers per second. But to make sense of the notions of space and time, it is necessary to imagine that those there-and-now events are somehow really "out there," spanning days, months, years and, by extension (you can magnify the mischief by increasing your changes in speed and the distance to "there"), *all* of time.

The idea that events in time are laid out "all at once" motivated

Einstein to write the words quoted at the start of this section. But this concept by no means originated with the theory of relativity; it recaptures a faint echo of the notion of eternity robbed from Western humanity by Newton. Its deep fascination for writers and poets is explicitly captured by the words of William Blake: "I see the Past, Present and Future existing all at once, before me,"[12] and eloquently echoed in the lines of T. S. Eliot:

> And the end and the beginning were always there
> Before the beginning and after the end.
> And all is always now.[13]

It did, however, require something of the power and testability of the theory of relativity to force scientists into a radical reappraisal of their conception of time—in particular, to get away from the notion of "things happening" in an orderly and universal sequence, and to start regarding time, like space, as simply "there." Just as we can survey space as a landscape spread before us, so we can survey time (in our mind's eye, at least) as a *timescape* timelessly laid out. Philosophers refer to the timescape concept as "block time," to distinguish it from psychological (and commonsense) ideas of "the fleeting present."

Block time suggests we represent time after the fashion of space. The first physicist to suggest this was Hermann Minkowski, who had been one of Einstein's teachers at the ETH. In 1908, Minkowski gave a lecture in Cologne on the subject of his former student's remarkable new theory of relativity, commencing with the dramatic statement: "Henceforth space by itself, and time by itself, are doomed to fade away into mere shadows, and only a kind of union of the two will preserve an independent reality."[14]

The "union" to which Minkowski alluded was his idea. If time can be spatialized, at least for the purposes of mathematical representation, then it must be treated as a *fourth* dimension, because there are already three dimensions of space. This sounds rather arcane, but the spatialization of time has been going on for as long as mankind has used symbolic representation. Writer Anthony Aveni points out in his fascinating book *Empires of Time* that our paleolithic ancestors were denoting intervals of time by sequential notches on bones at least twenty thousand years ago, and this is surely a spatial representation of time. Even the terminology "the fourth dimension" was used to describe time years before the theory of relativity burst on the scene. In his 1880 essay "What is the fourth dimension?" the British scientist Charles Hinton invited us to imagine "some stupendous whole, wherein all that have ever come into being or will come co-exists," which is a fairly vivid description of block time.[15] Furthermore, this arrangement "leaves in this flickering consciousness of ours, limited to a narrow space and a single moment, a tumultuous record of changes and vicissitudes that are but to us."[16] In other words, Hinton asserts that the now of our conscious awareness is merely a subjective phenomenon—of which, more later.

What was new about Einstein's time was the fact that it connected time to space *physically,* not just metaphorically. The theory of relativity interweaves space and time in a rather precise and intimate way. I have mentioned how space shrinks as time expands. Mathematically these distortions are convolved in the same set of formulae. Minkowski emphasized that he was not simply tacking an extra time dimension onto the three space dimensions for fun, but because the resulting entity formed a *unified* "spacetime continuum," in which the purely spatial and the purely temporal aspects could no longer be untangled. The theory of relativity does not permit us to separate time from space by taking spatial, or equal-moment, slices through spacetime in an absolute and universal way. Each observer will have his or her particular slicing, but they will not in general agree. A picture of spacetime might be helpful at this stage. Fig. 2.2 shows what are known as Minkowski diagrams, depicting space and time together. One of the problems with these diagrams is that it is not possible to draw four dimensions on a sheet of paper, so at least one space dimension has to be left out. Space is represented horizontally while time runs vertically. The diagram shows how different observers slice up spacetime into "space" and "time" in different ways.

Hermann Weyl, a close associate of Einstein's, expressed the new "spacetime" view as follows:

> The scene of action of reality is . . . a four-dimensional world in which space and time are linked together indissolubly. However deep the chasm that separates the intuitive nature of space from that of time in our experience, nothing of this qualitative difference enters into the objective world which physics endeavors to crystalise out of direct experience. It is a four-dimensional continuum, which is neither "time" nor "space."[17]

Einstein himself wasn't too thrilled with the unified spacetime idea at first, dismissing Minkowski's new four-dimensional geometry as "superfluous" pedantry, but he came around to the idea in due course. The true significance of this unified four-dimensional *spacetime* is that it possesses a common geometry which thoroughly mixes up the space bits and the time bits. Minkowski was quick to work out the rules of spacetime geometry. Unfortunately, they are not a straightforward generalization of three-dimensional school geometry extended to accommodate an extra dimension (though they are not hard either). I shall give more details in Chapter 8, but for the purposes of the present discussion I wish

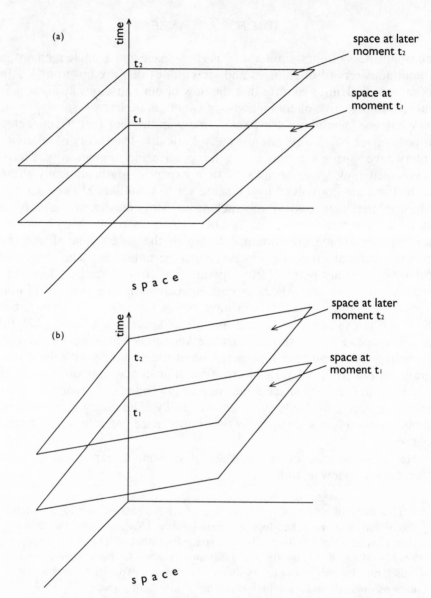

2.2 *Spacetime, according to Einstein's theory of relativity. In these so-called Minkowski diagrams, time is drawn vertically, and two dimensions of space are drawn horizontally. The horizontal slices in (a) denote space at the two moments* t_1 *and* t_2*, as seen by a particular observer. All points on a given slice are judged to be simultaneous by that observer. In (b) the same spacetime is sliced differently, corresponding to the perspective of a second observer, in motion relative to the first. It is the points on the sloping planes that are regarded as simultaneous by the second observer. Thus there is no unique and agreed way of slicing up spacetime into "space" and "time." Method (a) may look more natural, because the slices are horizontal, but that is only because I have arbitrarily drawn the axes of the diagram to correspond to the reference frame of the first observer.*

merely to note that when we draw diagrams in spacetime (as opposed to space) our normal intuition about distances and angles can lead us astray.

Minkowski diagrams can still be a great help, however. Let me illustrate the experiences of Ann and Betty using one (Fig. 2.3). For artistic convenience I have retained only one space dimension. First note that an event, such as Betty leaving Earth, corresponds to a single point in spacetime. An object, such as a person or a rocket, traces out a path in spacetime called its "world line." Ann's world line, which coincides with that of the Earth, is simply a straight line. The line is vertical because I have decided to draw this diagram to represent events as observed in the reference frame of the Earth. In this frame Ann doesn't move, so as time "goes on" she just traces out a line with fixed spatial coordinates. By contrast, Betty races off in the rocket, along a world line that tips away to the right, then reverses, and goes back to Earth again. The events representing Betty's departure from Earth, arrival at the star and return to Earth are labeled P, Q and R respectively.

Now, the crucial point is this. The duration between the two events P and R is not fixed, but depends on the *length* of the world line that the observer follows between them. The drawing makes it obvious that the

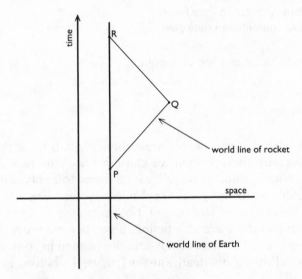

2.3 Minkowski diagram of the twins effect. Betty leaves Earth at event P. The sloping line is the "world line" of her rocket, which reaches the distant star at event Q. The abrupt change in slope of the world line at Q represents Betty's about-turn. Reunion on Earth occurs at event R. The distances along the alternative spacetime paths PR and PQR are obviously different, indicating different experienced durations between the events P and R. In fact, because of the strange rules of Minkowskian geometry, PQR is the shorter route.

distance between P and R as measured along the world lines will be different: Ann has a straight world line but Betty's is kinked via Q. You might think this would make Betty's estimate of the duration longer, but here the diagram fools you, as I warned. The wonky geometry of Minkowski space differs from "normal" geometry here, in that lines sloping away from the vertical must be multiplied by a special foreshortening factor. When this is done, it turns out that the *longest* time between two events is actually the time told on a clock that has a *straight* world line connecting the events. So Betty "gets to" event R in a shorter time than Ann. Notice I don't say "Betty gets to R first," because R isn't a place but an event. An event involving both Ann and Betty (in this case their reunion) can't be experienced at different moments, even though Ann and Betty disagree about the respective *durations* since the time of departure (event P).

Spatializing time like this may advance our understanding of physics, but a heavy price has been paid. Human life revolves around the division of time into past, present and future; people will not relinquish these categories just because physicists say they are discredited. T. S. Eliot concurred poetically with Minkowski:

Time present and time past
Are both perhaps present in time future,
And time future contained in time past,

But he went on to point out the consequence:

If all time is eternally present
All time is unredeemable.[18]

This is perhaps what disturbs people most about block time. If the future is somehow "already there," then we can have no hand in shaping it. The proverb "What's done is done" would seem to apply with equal force to the future as to the past. Weyl once wrote: "The world does not happen, it simply is."[19] Happening, becoming, the flux of time, the unfolding of events—these are all a fiction, if you believe Weyl. Einstein did; hence the quote at the start of this section, penned in consolation to Besso's widow following his death (just a few weeks before Einstein's own).

In their professional lives most physicists accept without question the concept of the timescape, but away from work they act like everybody else, basing their thoughts and actions on the assumption of a moving present moment. For can anyone really be convinced that the future won't *happen*, but somehow will simply be (when its time comes . . .)? That any impression to the contrary is some sort of delusion? Suppose you had a medical condition that required surgery, and you were informed by your physician that an anesthetic would be dangerous. Would you consent to go ahead anyway without anesthesia, on the basis that, once the operation was over, the pain would be "merely a memory"? Probably not. However, there is a drug called midazolam which has the effect of erasing short-term memory. If a patient wakes accidentally during surgery and experiences great pain, the drug ensures that, afterwards, she or he is blissfully ignorant of the horrible experience; to the postoperative patient, it is as if the anesthetic had worked perfectly. The question is, suppose you were offered that drug in place of an anesthetic —would you accept it? Or would you prefer to risk the anesthetic, on the basis that the experience of pain would be real at the time (even if you had no recollection of it afterwards) and that the painful experience still lay in the future—it would not "have happened" yet? I know what my answer would be.

Even Einstein confessed, near the end of his days, that the problem of the now "worried him seriously." In conversation with the philosopher Rudolf Carnap he conceded that there is "something essential about the now," but expressed the belief that, whatever it was, it lay "just outside the realm of science."[20] Maybe, maybe not. That is all I want to say about the subject of now for now. But I shall return to it later . . .

CHAPTER 3
TIMEWARPS

Is Einstein's theory a crazy vagary? It surely is.

A 1921 EDITION OF *THE NEW YORK TIMES*

THE LIGHT BARRIER

One of the most desolate places on Earth lies about five hundred kilometers north of Adelaide in South Australia. The land is desert, but not the drifting yellow sand dunes of the Sahara. Here the soil is a rich red, the terrain almost completely flat, and the ground decorated by desperate-looking bushes that somehow manage to cling to life in the hot and arid conditions.

It seems odd to find a town here; water is piped all the way from the river Murray, a distance of several hundred kilometers. The name of the town is Woomera, which means "throwing stick" in aboriginal Australian. I went there to see with my own eyes, and a theoretician's childlike curiosity, the experimental evidence for one of the world's biggest measured timewarps. At least, I had come to see the equipment. More accurately, I had come to watch the latest upgrade of the equipment being "unveiled" by the Minister for Science. The new system rejoices in the corny acronym of CANGAROO, standing for Collaboration between Australia and Nippon for Gamma Ray Observations in the Outback.

The scientific station is not located in the town, but a short drive down a sealed road through the scrub, near the place where they used to launch rockets. Not many people know that Australia was the fourth power in space (France just beat Australia to the number-three place). A satellite made in Adelaide was launched into orbit atop an off-the-shelf

U.S. rocket in 1967. At one stage the Woomera base was testing or launching dozens of rockets a year, mainly for Britain, then for her European partners. The funding stopped in the early 1970s, and the Australian government, evidently of the opinion that there was no future in space technology, ordered the facilities to be blown up. There is still a military base at Woomera, and one day rockets may fly again, but for now the main scientific activity there concerns cosmic radiation. This is what CANGAROO is designed to study.

The existence of a form of penetrating radiation from space was first suspected over a century ago, and has been a source of wonder and discovery ever since. Several new subatomic particles have, over the years, first been identified amid the detritus of cosmic rays. As I explained in the previous chapter, the particles experienced near the surface of the Earth are subatomic fragments produced when high-speed particles from space (which turn out to be mainly protons) strike atomic nuclei in the atmosphere.

The Woomera equipment targets cosmic rays in a clever way. When a high-energy primary particle strikes, it creates a shower of secondary particles themselves having pretty enormous energies. The shower is directed downwards by the momentum of the incoming primary particle, and fans out a bit before it hits the ground. Some of the electrically charged particles involved in these showers move very close indeed to the speed of light. In fact, they actually move faster than light moves through the air. This is an important point. The theory of relativity forbids a subatomic particle from going faster than the speed of light in a vacuum. But light travels more slowly in air, so it is possible for a nuclear particle, which may be slowed only slightly by the air, to be superluminal in the atmosphere. If the particle is electrically charged, it creates a sort of electromagnetic shock wave, a little bit like a sonic boom, but with light instead of sound. The light is known as "Cherenkov radiation," after its Russian discoverer. Cherenkov radiation is easy to identify by the angle of its beam, so the Woomera scientists have cleverly built a device to do just that.

The system works by scanning the dark night sky, and registering the tiny Cherenkov flashes that betray the passage of a cosmic-ray shower. I can't resist telling the story of an earlier system, located at Buckland Park, much closer to Adelaide and, like CANGAROO, designed to study cosmic-ray air showers. The principal investigator was Roger Clay, a talented experimental physicist and lifelong trombone player who has spent most of his career investigating cosmic rays. In 1974, Clay and his colleagues were excited by some unusual data from Buckland Park. Taken at face value, the information seemed to suggest that some of the particles in the air showers were not only reaching the ground ahead of light, but actually traveling faster than the speed of light *in vacuo*.

Now, this was sensational stuff. As I have emphasized, the theory of relativity forbids any particles from breaking the light barrier. Were they to do so, the consequences for the nature of time would be profound. There is even a limerick that warns us:

There was a young lady named Bright
Whose speed was far faster than light;
She set out one day, in a relative way,
And returned on the previous night.

In short, faster-than-light can mean backwards in time, with all the puzzles and paradoxes that follow (see Chapter 10).

In fact, the theory of relativity does not say that "nothing can go faster than light," as is often reported. It does permit objects to travel at superluminal speeds, even in a vacuum, but only if such objects can never travel *slower* than light. In other words, according to Einstein's theory, nothing can *cross* the light barrier, by going either up or down in speed. Physicists have invented a name for superluminal particles: they are called "tachyons," after the Latin word for "speed." Roger Clay and his colleagues believed they had found tachyons.

Although tachyons are not actually ruled out by the theory of relativity, they are an embarrassment to physicists, not least because they could be used to send signals into the past. (Ms. Bright cannot travel bodily into the past, in the manner just described, without violating the theory of relativity, but she could perhaps manipulate tachyons to send a message backwards in time. When it comes to the paradoxes of time travel, this is more or less as bad, as we shall see.) There are other problems too, of a more technical nature, about incorporating tachyons into the framework of current physical theory. If you were to conduct a poll among physicists, I guess you would find about 90 percent against the idea of tachyons, 1 percent in favor and the rest "don't knows." (Remarkably, Lucretius suggested the possibility of faster-than-light particles, though he was unaware of their temporal implications.) I well remember the ballyhoo that resulted when the Australian group announced their discovery of, maybe, tachyons. It would have been exciting. But a more cautious appraisal of the data led them to downplay the claim, and turn their attention to other things.

Among these other things are some of the fastest known ordinary particles (i.e., not tachyons) in the universe. Physicists like to describe high-speed particles by their energy rather than their speed. This is because, as the speed of light constitutes a barrier, all very fast particles move at more or less the same speed—just a little bit less than the speed of light *in vacuo*. So one particle may have ten times the kinetic energy of a similar particle, but move only a snail's pace faster. The energy method of reckoning is more natural when it comes to discussing time dilation too.

To elaborate this point, let me put in some numbers. Particle energies are measured in a curious unit called an "electron volt." This is the energy that one electron would acquire if it was accelerated in an electric field through one volt of potential difference. To get a feel for it, note that the typical kinetic energy of an electron circling within an atom is just a few electron volts. By comparison, the typical energy of a primary cosmic-ray particle would be a trillion electron volts, which suggests that there are cosmic dynamos out there somewhere generating at least a trillion volts. Most cosmic-ray primaries are protons. A proton with a trillion electron volts of kinetic energy moves at about 99.9999 percent of the speed of light, whereas a proton with energy of ten trillion electron volts moves at about 99.999999 percent of the speed of light. At these speeds, it is more informative to give the *difference* in speed between the proton and that of light. For a ten-trillion-electron-volt particle it is just three meters per second—about walking pace. At a hundred trillion, the difference is only three centimeters per second—a literal snail's pace—and at a thousand trillion it drops to a mere 0.3 millimeters per second. And so on. Notice how further increases in energy imply an ever-diminishing advance in speed. (While at Woomera, I couldn't help remembering that a little farther to the north lies Lake Eyre, a mostly dried-up basin in which the British adventurer-playboy Donald Campbell broke the land-speed record in 1964. He achieved just 429 miles per hour—691 kilometers per hour—or 0.6 millionths of the speed of light.)

To convert cosmic-ray energies into a time-dilation factor, you use a simple formula: divide the energy of the proton in electron volts by a billion. This gives the stretching factor for the clock rate. Thus a trillion-electron-volt proton has its time slowed to one-thousandth of our rate, while for a thousand-trillion-electron-volt particle the factor is one-millionth.

The CANGAROO facility seeks out air showers produced not by protons but by gamma-ray *photons* with energies in the range of one trillion to ten trillion electron volts. (A photon is a packet, or quantum, of light. Gamma rays are very short-wavelength photons.) Even these enormous energies are modest by cosmic-ray standards. In 1993, a primary cosmic ray (almost certainly a proton) with an energy of three hundred million trillion electron volts was spotted by an American group using a facility called the Fly's Eye. The curious name derives from the optical geometry employed. The Fly's Eye detector system consists of more than a hundred 1.5-meter-diameter mirrors oriented in many different directions, like the components of the compound eye of a fly. With this arrangement most of the night sky can be covered at once. The array sits on an arid bluff almost as desolate as Woomera, overlooking a ballistic missile silo in western Utah. But the Fly's Eye hunts nuclear particles rather than nuclear weapons, tracking cosmic-ray primaries with the highest known energies.

At a hundred million trillion electron volts, a single proton packs the same punch as a pitched baseball, and the timewarp factor is a staggering one hundred billion. A clock moving alongside such a particle would appear to us to tick at one-hundred-billionth of the rate of the clock on my office wall. Every day that passes on Earth corresponds to just one microsecond of particle-time (and vice versa, of course). A tick of an office clock pacing such a cosmic ray would occur but once every three thousand Earth years. This enormous warp factor has implications for the nature of the cosmic-ray particles involved. Actually, nobody knows precisely what produces cosmic rays, especially those with energies as high as a hundred million trillion electron volts. Supernovae, exploding galactic cores, pulsars and black holes are all possible cosmic-ray sources, but no simple mechanism seems to explain all the high-energy particles coming from space. Part of the problem is that cosmic rays pepper the Earth more or less evenly from all directions, so it is hard to identify specific sources. Also, charged particles like protons are deflected by the galaxy's magnetic field, so the direction of their arrival may not give much information about their origin.

One exception to this is an object known as Cygnus X-3, an X-ray source consisting of a pair of imploded stars situated thirty-five thousand light-years away in the constellation of Cygnus. In the mid-1980s, the Fly's Eye and other detector systems hinted at energetic cosmic rays coming from the direction of Cygnus X-3 in a straight line. To avoid magnetic deflection, the particles had to be uncharged; this rules out protons. Physicists began wondering whether some exotic new type of electrically neutral particle might be involved. Some theories predict the existence of heavy neutral particles called "photinos." Could the cosmic rays from Cygnus X-3 be photinos? Perhaps. But there was another exotic possibility. The humble neutron is uncharged. Might this be the mystery-particle? Neutrons do not normally figure in cosmic-ray studies, because they are unstable. The half-life for a neutron to decay is about fifteen minutes, and you can't travel far in that time. But this is where the timewarp comes in. If the neutron moves fast enough, then, in our frame of reference, its lifetime could become enormously extended. At a million trillion electron volts and a warp factor of a billion, fifteen minutes translates into thirty thousand years. This means such a neutron could travel thirty thousand light-years across space before decaying, more than sufficient to get to Earth from Cygnus X-3. This is time dilation with a vengeance! If you could travel that fast, you would live for billions of Earth years.

Does this mean high speed is the secret of eternal youth?

No! Many people fall for this. The above remark about living for billions of years means: In the frame of reference of Earth, your life span of seventy-five years occupies billions of years of Earth time. In your own frame of reference, seventy-five years remains seventy-five years. From your perspective, it is the events on Earth that are slowed. One tick of a clock on Earth would correspond to three thousand of your years. Unfortunately, you can't use relativistic time dilation to delay your own aging process relative to your own experience of time, only relative to someone else's.

Now I'm confused about something, complains our skeptic. I keep reading that the age of the universe is fifteen billion years. But whose fifteen billion years is that? If some cosmic rays can telescope trillions of years into seventy-five, doesn't that mean the universe began about a year ago, cosmic-ray time? Or maybe I got that around the wrong way, and it's one year of our time that equals billions of years cosmic-ray time. Surely they see our fifteen billion years stretched to a billion trillion years? Come to think of it, isn't the universe full of motion, with galaxies rushing apart, some of them near the speed of light? Surely Einstein's flexitime makes a nonsense of dating the origin of the universe? It doesn't really have *a date, does it?*

Well, yes and no. We shall see in a later chapter how to unravel these different times. But the objection is a valid one. Once we accept that time is no longer absolute and universal, the question whether there exists some sort of cosmic time, and whether it is unique, is crucial. We don't have intuition to guide us here, because in daily life time gives a convincing performance of being the absolute and universal dimension we know it isn't.

The reason flexitime isn't part of our everyday commonsense experience is that human beings rarely achieve relative speeds greater than a millionth of the speed of light, and any time dilation is too small to notice. In 1905, the train was the fastest form of transport, and early discussions of relativistic time often referred to observers on trains. However, Einstein did make use of the fact that the Earth rotates faster than any train, concluding that "a balance clock at the equator must go more slowly, by a very small amount, than a precisely similar clock situated at one of the poles under otherwise identical conditions."[1] Although Einstein didn't know it at when he wrote the above words, the Earth is actually responsible for *two* timewarps, which actually cancel— so he was wrong! One is due to the Earth's rotation, the other to its gravity. It was Einstein himself who discovered the gravitational effect on time, a couple of years later.

Why does gravity affect time? There are many fascinating arguments that suggest it must. One of them has to do with that old engineer's dream: perpetual motion.

PERPETUAL MOTION AND THE UPHILL STRUGGLE

There's no such thing as a free lunch.

MILTON FRIEDMAN

If papers on time top the list of crank manuscripts circulated to university physics departments, those on perpetual motion come a close second. The search for a machine that would give something for nothing has a long and dismal history, stretching back to antiquity. In its time it engaged the attention of such luminaries as Leonardo da Vinci and Robert Boyle. Writing in 1906, the Assistant Examiner of the British Patent Office noted that his organization had received six hundred patent applications for perpetual-motion machines since 1617.[2] He went on to describe one of the more common proposals, which I have sketched in Fig. 3.1. The machine consists of a continuous conveyor belt with cups attached. On one side of the device the cups are filled with balls; on the other side they are empty. The weight of the balls pulls the belt down on the former side. When the balls reach the bottom, they roll out of the cups and into a giant corkscrew mechanism, which conveys them to the top again. The motive power to turn the corkscrew comes from the revolving belt itself, communicated via a system of gears. We are asked to believe that the entire contraption will keep turning without needing an engine to drive it—indeed, that it may even deliver surplus energy for free.

There are two laws of physics that tell us the quest for such perpetual motion is doomed to failure. The law of conservation of energy is the first. It says that in a closed system you can't get out more energy than you put in; all you can do is move energy around or change its form and hope you don't lose control of too much of it in the process. What slips through your fingers eventually appears in the form of heat. When all the motive energy is turned into heat, the machine stops. The law of conservation of energy is sometimes called the first law of thermodynamics. The second law of thermodynamics tells you that, when energy has dissipated away as heat, you can't get it back again without using up at least as much energy in the process. Anyone trying to generate energy without fuel will run foul of the first law. Anyone merely wanting perpet-

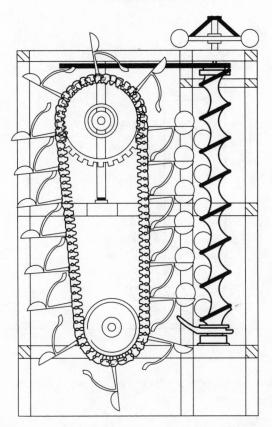

3.1 Something for nothing? The picture shows the design of a typical early perpetual-motion machine.

ual motion will run foul of both the first and the second laws, because in any real system there is always dissipation—usually in the form of friction—that slowly saps the energy of any moving system. That is why all motors need a supply of fuel to keep running. In the case of the conveyor belt, the energy delivered by the falling balls would not be enough to pay for the corkscrew action *and* the inevitable heat produced as the gears grind around. The rising balls would stop short of their destination.

Fig. 3.2 shows a design for a perpetual-motion machine due to the mathematician Hermann Bondi, said to be based on an idea by Einstein himself, and clearly inspired by the ancient design of Fig. 3.1. It consists once again of a conveyor belt with a lot of cups attached at regular intervals. Each cup contains, however, not a ball, but a single atom. The atoms in the left-hand cups are in excited states, whereas those on the right are in their lowest-energy, or ground, states. The principle of the machine depends on Einstein's formula $E = mc^2$. This says that energy E has mass m, and because mass has weight, we can deduce that the

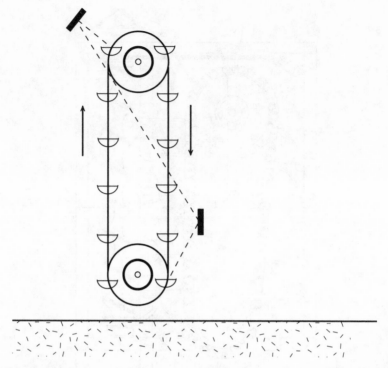

3.2 Gravitational timewarping prevents this ingenious device, due to Hermann Bondi, from creating energy.

excited atoms (which have more energy than those in their ground states) weigh more. I should make it clear that this is not a practical proposal in the form described, because the extra weight of the atoms is minuscule, but the setup is intended as an idealization to illustrate a point of principle.

The lopsided weight distribution causes the belt to turn, as the heavier left-hand atoms pull the belt down on that side. At the bottom of the conveyor is some sort of gizmo to induce the excited atoms to give up their energy in the form of photons of light. Induced photo-emission by atoms is familiar to physicists and occurs in lasers, for example. This arrangement ensures that when the cups reach the right-hand part of the conveyor their atoms are in the lighter ground state, as required. The emitted photons are directed towards the top of the conveyor—they travel there without the assistance of a corkscrew—where they are reflected into the cups arriving there and used to excite the atoms contained therein. In this way the atoms on the left are kept excited and those on the right are kept de-excited as the conveyor continues turning. Thus the lopsided conditions are maintained and the motion could be used to generate electricity, apparently *ad infinitum,* without using any fuel.

Now, Einstein and Bondi never supposed that they had hit upon a loophole in the laws of thermodynamics. Instead, they accepted that perpetual motion is impossible, and proceeded to argue the other way around. The assumptions built into the conveyor design, it seems, must be unphysical. There has to be some reason why the photon energy liberated at the bottom is not enough to pay for the excitation of the atoms at the top. The deficit should be precisely equal to the energy which the machine supposedly creates. That way the books will balance.

The energy of the turning conveyor comes from gravity, which pulls the heavier atoms downwards. This suggests that gravity does something compensatory to the climbing photons to rob them of the energy gained when the atoms descend. Evidently, as the photons struggle uphill against gravity, they are weakened. As a result, they arrive at the top with less energy than they started out with at the bottom, and so are unable to excite the atoms to the same levels as before. The conveyor will then gradually slow down and stop—the ignominious fate of all *perpetuum mobiles*.

The conveyor device hints that gravity affects light, but what has that got to do with time?

WHY TIME RUNS FASTER IN SPACE

Before I explain the connection between light weakening and time, let me describe another argument that confirms the link between light and gravity. Einstein himself arrived at the idea that gravity affects light by an entirely different line of reasoning, in 1907. By this stage of his career, his genius had been recognized by the scientific community, but he was still employed at the Swiss Patent Office; he did not yet have a university position. (History does not record whether he encountered perpetual-motion patent applicants similar to those of his British counterparts.) He had, however, been awarded a Ph.D. by the University of Zurich for a research project in statistical mechanics.

It seems his employment duties were undemanding enough for Einstein to have plenty of time to sit and think about the nature of the physical universe. In 1907, whilst contemplating the mysteries of gravitation, he came up with an elegant line of reasoning that typifies the way in which he was able to draw profound conclusions about the world based on pure thought. Like Galileo three centuries earlier, Einstein began his deliberations on gravitation by comparing a gravitational force to an acceleration.

First he imagined what an acceleration felt like. In the final version of the argument, he used the example of an elevator suddenly starting up. We are all familiar with the way that accelerated motion produces "g forces"—that is, it feels just like gravity. An upwardly accelerating elevator presses you towards the floor, adding to your weight, whereas a downwardly accelerating elevator "leaves your stomach behind" as it temporarily reduces your weight. Another example of acceleration mimicking gravity is rotation. In Stanley Kubrick's film *2001: A Space Odyssey,* the space station is shaped like a wheel, and slowly turns to create "artificial gravity" along its rim. Although Galileo and Newton were aware of the close link between acceleration and gravitation, they regarded it as an incidental feature of nature. Einstein elevated it to a fundamental principle, which he called the "principle of equivalence," which states that, in the immediate proximity of an accelerating system, the acceleration is physically equivalent to a gravitational force.

The next step of Einstein's argument was to note that motion has an effect on light, the Doppler effect that I mentioned in connection with the twins effect in the last chapter. A good example of the *auditory* Doppler effect at work is when a police car with siren blaring rushes by. The pitch of the siren falls suddenly as the car passes (wee-wee-wee-wee-wow-wow-wow . . .). This happens because the onrushing car compresses the sound waves before it, boosting their frequency. Conversely, when the car is receding, the waves coming back at you get stretched to a lower frequency. The same thing happens with light waves: light from an approaching source suffers a frequency rise, whereas light from a receding source suffers a frequency fall (only a very small change for everyday speeds). Because the frequency of light is related to its color, the Doppler shift for light amounts to a color shift. The long-wavelength end of the visible spectrum is red, the short wavelength is blue, so an approaching source is blue-shifted, a receding source red-shifted. The Doppler effect applies to all electromagnetic waves; it is, for example, employed in police radar traps to spot speeding motorists.

By splicing the principle of equivalence and the Doppler effect, Einstein ingeniously deduced that gravity affects light. Imagine accelerating away from a source of light. As your speed rises, so the light will become more red-shifted by the Doppler effect. Therefore, reasoned Einstein, it should also become red-shifted by a gravitational field, because an acceleration mimics a gravitational field and should produce equivalent physical effects. Using his special theory of relativity, he was able to arrive at the formula that describes the magnitude of the gravitational red-shift effect.

It is this red shift which saves us from the *perpetuum mobile* paradox, because there is a relationship between the frequency of light and the energy of the corresponding photons. In fact, these two quantities are in direct proportion. Thus, if light is red-shifted, the photon energy is reduced; so, in the conveyor system, the photons arriving at the top of the device will indeed be enfeebled, and unable to excite the atoms there.

We are now ready to make the all-important connection with time.

The word "frequency" means number of cycles per second; if the frequency of light goes down as a result of a gravitational red shift, the number of cycles of the wave that pass each point in space per second is reduced. But to measure frequency we need to use a *clock* to count the seconds. So, if light from the bottom of the conveyor arrives at the top with a lower frequency, we could say either that the light frequency had been reduced, or, equally well, that time at the bottom of the conveyor runs a bit slower than time at the top. After all, since we can only measure frequency using clocks, a change in frequency is equivalent to a change in clock rates. Isn't it?

That seems a bit of a cheat, complains our ever-vigilant skeptic. Why can't we simply say the frequency of light changes with height, and maintain that *time* is the same at all heights?

Well, suppose we used the cycles of the light wave as the beats of a clock? It would make a perfectly good clock. In that case, the gravitational red shift would amount, directly, to a change in clock rate.

Okay. But what if we use another sort of clock? We don't have to use a light-wave clock. You can't claim that *time itself* changes with height unless *all* clocks are affected in the same way.

Indeed so. And they are! Here is the reason why. The tick-tocking activity of a clock has some associated energy, and this possesses weight, like all forms of energy ($E = mc^2$ again). If you lift a clock to a greater height, you have to do work on it—against its weight, so to speak. The work done appears as gravitational energy stored in the clock; you could get it back by allowing the clock to fall back down. Now, a very small part of the total weight of the clock comes from its internal energy—the tick-tock energy. Therefore, a portion of the extra energy gained by the clock when it is raised up comes as a result of our lifting the tick-tock weight. This portion (tiny though it is) shows up in the guise of extra tick-tock energy, as a result of which the clock ticks a bit faster. So raising a clock makes it run faster! A careful study reveals that the clock's rate changes with height in exactly the same way as a light wave or a photon loses frequency as it climbs. Moreover, the effect is independent of the design of the clock. Whatever clock you are interested in (including the human brain), it will run faster up there than down here. And the change in rate is identical for every sort of clock. So, rather than saying, "All clocks run faster up there," it is better to say, "Time runs faster up there."

Let's take stock of the reasoning so far. We have been led to conclude that, because perpetual-motion machines are impossible, time "speeds up" with height. Einstein arrived at the same conclusion from a consideration of accelerating elevators and the Doppler effect. Both arguments suggest that, the higher up you go, the faster time runs. Apart from mentioning that the effect is tiny, I haven't given any figures, but, to take one example, a clock on the ground will, after an hour, lose a nanosecond (one-billionth of a second) relative to a clock in space. The effect also implies that time runs slightly faster at the top of a building than at the bottom. Over a lifetime, you could gain a microsecond or so over your high-rise neighbors simply by living on the ground floor. You might be inclined to think that such tiny temporal distortions are both undetectable and utterly insignificant. In fact, they are neither. Not only can they be measured, but under some circumstances gravitational timewarps can grow enormous and lead to dramatic effects, as I shall shortly discuss.

You would be right to be skeptical of the foregoing theoretical arguments taken alone. If time really does change with height, it is important to demonstrate this experimentally. Before coming on to that, I'd like to give a final argument for the effect of gravity on time. Ironically, this third argument was used *against* Einstein—who had overlooked it himself—in a famous debate with the Danish physicist Niels Bohr. Their encounter took place much later—in 1930—by which time Einstein was an international superstar professor with a Nobel Prize. But Bohr's intellect was every bit a match for him.

THE CLOCK IN THE BOX

It happened that Bohr and Einstein spent years at loggerheads. The bone of contention was not the theory of relativity, which was rapidly accepted by the physics community, but the equally revolutionary and equally disturbing quantum theory. Recall that Einstein had a hand in the inception of this theory when he gave a successful explanation for the photoelectric effect in 1905, the same year he published his first paper on relativity. It was not until the late 1920s, however, that quantum physics was placed on a solid foundation in the form of a comprehensive "quantum mechanics," an enterprise presided over by the great Niels Bohr.

One of the founders of the new quantum mechanics was the young German physicist Werner Heisenberg. In 1927, he proposed a fundamental principle of quantum physics. Known as the "uncertainty principle," it sets strict limits on the degree of precision with which we can determine the properties of a particle. For definiteness, think of an electron. Roughly speaking, all measurable attributes of the electron are subject to uncertainties in their values. So, for example, you might like to know where the electron is located and how fast it is moving. Heisenberg's uncertainty principle tells you that you cannot accurately determine *both* these quantities at the same time. The more precisely you measure the position of the electron, the less precisely you can know its motion, and vice versa. There is an inescapable tradeoff between these two variables. If you know where an electron is, you are vague about its motion; if you know how it is moving, you can't know exactly where it is. Similar uncertainty relations apply to other pairs of quantities. An important case concerns the energy of the particle and the time at which the energy is measured; they are also mutually uncertain.

These fundamental uncertainties cannot be derived from normal or commonsense ideas about particles—i.e., from what is known as classical physics. They are entirely a property of the quantum world. The vagueness or fuzziness that the uncertainty principle expresses is closely associated with another sort of vagueness, known as "wave-particle duality." An entity such as an electron, which we normally think of as a particle, sometimes takes on the features of a wave. Conversely, light, which we normally think of as a wave, can behave like a stream of particles (photons). Obviously, in the everyday world, something cannot be both a wave and a particle: they are quite different animals. But in the quantum domain, such a dual nature is possible, and either the wave or the particle aspect of the quantum entity will manifest itself, depending on the circumstances. You must not try to imagine what, say, a photon "really" is, for that question is almost certainly meaningless. It is like nothing that we can encounter in the macroscopic world of human experience.

I mentioned in the previous section that the energy of a photon is proportional to the frequency of light. This statement sounds innocuous enough, but it conceals a subtlety. The concept of frequency makes sense only when applied to a wave, whereas the energy of the photon refers to a particle. So we have wave-particle duality at work here. Obviously, a measurement of the frequency of a wave takes time—you have to let the wave go through a few cycles and measure their duration. If you try to chop a light wave up into tiny chunks of very short duration, you will no longer have a wave with a single well-defined frequency, or photons with a single well-defined energy. So you can see the Heisenberg tradeoff operating here: To try and fix the photon's energy demands many cycles of the wave, which inevitably occupies a certain time. To try and determine where the photon is at a fixed time means chopping out a little chunk of wave and thereby smearing the energy. This is therefore one way to think about the energy-time uncertainty principle.

Heisenberg's uncertainty relations lead to conspicuous and dramatic effects on an atomic scale, but we do not notice them in everyday life: they are far too small. However, for consistency, these relations must apply to *all* physical systems, whatever their size or mass; otherwise we could use macroscopic objects to violate the uncertainty principle. So, even though we don't notice uncontrollable quantum changes in the positions or energies of ordinary bodies, they should nevertheless be present if the theory is to make sense.

Einstein didn't believe Heisenberg's uncertainty principle. More accurately, he did not feel that quantum physics, of which this principle is an indispensable part, gives a complete account of reality. He exercised great ingenuity in trying to find a flaw or a contradiction in the rules of quantum theory, only to have Bohr convincingly rebut him. By 1930, the Bohr-Einstein debate had reached new heights of sophistication. In that year, the French chemical company Solvay funded a conference in Brussels to discuss magnetism, and this gave Einstein an opportunity to present his latest argument against quantum mechanics for Bohr to chew on. At this stage, nearly all physicists were coming to accept the new quantum mechanics as an accurate and complete account of the world, weird though its conclusions might sometimes be. Einstein obstinately refused to follow the crowd.

In those prewar years, much of the discussion about atomic and subatomic processes remained purely theoretical. Certain key experiments had been performed, but the technology necessary to test thoroughly the conceptual foundations of the subject was not yet available. Physicists often spoke about, and drew diagrams of, quantum processes without seriously suggesting that the phenomena under discussion could be carried out in the laboratory. These were idealized "thought experiments," doable in principle but far too hard in practice. Many of them involved an analysis of quantum effects on macroscopic objects like metal screens and pulleys, effects that would be far too small for us to have any hope of measuring.

You might wonder how scientists can tell anything useful about the world merely from sitting around and thinking about utterly impracticable experiments. This raises interesting philosophical issues, and is worth a brief digression. Science is based on the assumption that the world is rational, and that human reasoning reflects, albeit in a somewhat shaky way, an underlying order in nature. Logical consistency requires that the various laws and principles which govern the natural world must fit together consistently. It is sometimes possible, by tenaciously following a logical thread, to make discoveries about the real world without ever conducting an experiment, simply by imagining a particular physical state of affairs. In practice, it is essential to confirm such theoretical predictions experimentally, as there are many examples in history of apparently rational thought producing absurd conclusions. Thought experiments can be both positive or negative: they may suggest new laws or principles, or show up inconsistencies in existing theories.

Einstein was a master of the thought experiment: "I believe that pure thought is competent enough to comprehend the world," he once wrote when explaining his reasoning processes.[3] I have already mentioned how he came to predict the gravitational timewarp effect by imagining how an acceleration, which feels like gravity, would produce a Doppler red shift in a light beam. He made no attempt to conduct an experiment of this sort, being content to rely on the mathematical and logical consistency of nature. In fact, thought experiments related to gravity have a long and honorable history stretching right back to Galileo, who first demonstrated that all falling bodies fall equally fast when air resistance is negligible. To prove it, Galileo experimented with falling bodies (legend has it, by dropping weights from the leaning tower of Pisa), but he also made use of a thought experiment. His intention was to refute the popular theory, due to Aristotle, that heavy bodies fall faster than light bodies. Imagine that a heavy body h is attached to a light body l by a slender string, and the two bodies are dropped from a tower (see Fig. 3.3). We can then ask the question: Does the presence of l increase or retard the rate of fall of h? Suppose Aristotle is right; then l should lag behind h. If so, the string will become taut and l will have

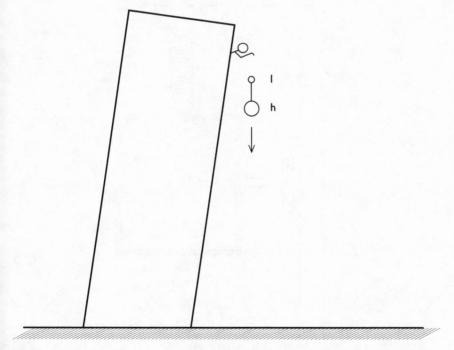

3.3 All bodies fall equally fast. The light body l, tethered to the heavy body h, can exert no influence on h's rate of fall, or there would be a logical contradiction.

the effect of *slowing* the fall of *h*. On the other hand, taken together, the system *l* plus *h* is heavier than *h* alone, and so should fall faster than *h* alone. If so, then *l* has the effect of *speeding* the fall of *h*. But this is plainly absurd: Aristotle's theory has led us to draw contradictory conclusions—namely, that *l* both decreases and increases the rate of fall of *h*. The only consistent account is that the presence of *l* has *no* effect on *h*: both *l* and *h* fall at the same rate. Thus Galileo was able to demolish Aristotle's theory without ever scaling the leaning tower of Pisa. In the same vein, Einstein set out to demolish quantum mechanics when he rose to speak at that famous conference in 1930.

The hypothetical apparatus for Einstein's thought experiment consisted of a metal box suspended from a rigid stand by a spring (see Fig. 3.4). The box has a small hole in the side. It also contains a clock, programmed to control a shutter which opens and closes across the hole for a brief duration at some preassigned time. The vertical position of the box can be measured by means of a pointer and scale. In addition to containing a clock, the box is filled with light. The experiment consists

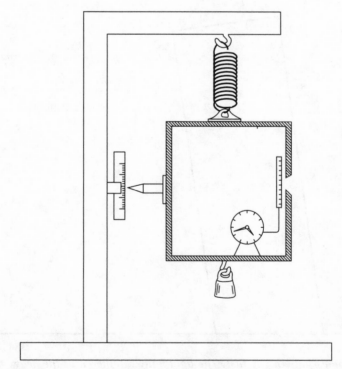

3.4 An artist's impression of Einstein's thought experiment to refute quantum mechanics. In fact, it demonstrates that time slows with height.

of an attempt to measure both the energy of a photon and the moment of its escape from the box, to unlimited precision, in blatant violation of Heisenberg's uncertainty principle. Here is what you do. First weigh the box. Then wait for the clock mechanism to open the shutter briefly, allowing a single photon to escape. The clock records the time when this happens. Now weigh the box again. Because it has lost a photon, it will be lighter. The difference in weight will yield the mass of the photon, and, by Einstein's formula $E = mc^2$, you can compute its energy.

The precision in the photon's energy is limited only by the accuracy of the weighing. You don't need actually to weigh the box before and after, as only the difference in weight is required. This may be inferred from the vertical movement of the box: After the photon has escaped it will be lighter, and so the pointer will indicate a higher position on the scale. By attaching a compensatory weight to the bottom of the box, the pointer can be returned, at rest, to its starting location. The value of this weight will then equal the weight of the lost photon. The duration the shutter is open can be made as brief as you like, so long as you have lots of photons inside the box. Thus the time of escape of the photon can be known very accurately too. Hence the accuracy of both the energy and the time seem to be limited only by incidental features that can in principle be arbitrarily refined. The weighty conclusion Einstein drew was that the uncertainty principle could in theory be circumvented by this procedure, and therefore cannot be a basic principle of nature.

Most of the younger physicists tended to shrug aside Einstein's repeated carping about quantum mechanics, with the remark "It'll be all right." On the other hand, Bohr the elder statesman always took Einstein's thought experiments very seriously, and on this occasion he was seriously rattled. Leon Rosenfeld, who attended the Solvay meeting, recalls that Bohr appeared to be in shock, and during dinner at the university club he spent his time trying to persuade colleagues that Einstein must have made an error. Bohr accompanied a serene Einstein back to their hotel in a state of great animation and agitation, following which he spent a sleepless night.

The next day, however, when it was Bohr's turn to speak, it was to be in triumph. To draw his conclusion, claimed Bohr, Einstein had overlooked his own theory of relativity! In measuring the energy of the photon, Bohr explained, you must accurately measure the vertical position of the pointer when it is at rest against the scale. To be sure that the pointer *is* at rest, you must determine that its vertical motion is zero. But this variable is *also* subject to Heisenberg's uncertainty principle: the more accurately you try to determine whether the pointer is at rest, the less certain you can be of its position—i.e., *where* it is at rest on the vertical scale. In other words, there will be an inevitable uncertainty in the *height* of the box.

Now, the essence of Einstein's experiment is that you can measure the energy of the photon by weighing it, so gravity plays a crucial role here. As we have seen, the theory of relativity predicts that higher clocks run faster than lower clocks, so if there is uncertainty in the height of a clock there will be a corresponding uncertainty in its rate. This translates into an uncertainty in the time at which the shutter opens to release the photon. Hence the very act of accurately measuring the energy of the photon inevitably introduces an uncontrollable error into the determination of the moment of its escape. All the effects involved are minute, of course, but Bohr put in the numbers and demonstrated that the familiar energy-time uncertainty relation emerges intact.

Poor Einstein was forced to concede his error, and in fact this decisive rebuttal by Bohr knocked the stuffing out of his extended crusade to discover something wrong with the foundations of quantum physics. The following year, he proposed Heisenberg for a Nobel Prize, with the endorsement: "I am convinced that this theory undoubtedly contains part of the ultimate truth."[4] But the whole truth? Perhaps not. To the end of his days, Einstein insisted that the quantum theory was, if not inconsistent, then certainly incomplete. It left out something crucial about reality, he believed. In spite of Einstein's formidable reputation as a scientist, his remained a minority view, and today, forty years after Einstein's death, refined experimentation has if anything made Einstein's position much harder to sustain. These days, very few physicists indeed doubt quantum mechanics, and the clock-in-the-box argument can be turned around. If quantum physics is to be consistent, then clock rates had better vary with height!

THE BEST CLOCK IN THE UNIVERSE

So much for thought experiments; what about real experiments? Do clocks really run slower in the basement? It's curious that much good physics takes place in basements. I used to suppose this was because university administrators held the subject in low esteem and gave the physicists the worst accommodation for their laboratories, but in fact the basement is often the best place to do accurate and sensitive measurements. Vibration is less: you can bolt things directly to the ground. Also, vehicle access is easier for heavy equipment.

One of my favorite basement labs is located in Perth, at the University of Western Australia. A patchwork quilt of rooms, corridors and halls of varying sizes and heights, it is crowded with an amazing jumble of pipes, wires, messy benches, computer terminals and empty Coke cans. In short, a typical physics lab. In a quiet corner of one of the research halls stands a row of intriguing blue cylinders about a meter and a half tall.

Some of them have sinister white vapor lazily leaking from gleaming steel valves at the top. One or two are boxed in by a bank of electronic gadgetry sporting light-emitting diode displays winking in dull red. On a nearby wall someone has pinned up a picture of Einstein. Presiding over this array of complicated equipment is a youthful-looking physicist by the name of David Blair. Australian by birth, Blair spends several months of the year globe-trotting, visiting other labs and attending conferences.

Now in his forties, Blair cuts an imposing figure. With a bushy beard and dark, penetrating eyes, a ready smile and a mop of unruly black hair, he could easily be taken for a farmer. In fact, he is a farmer of sorts. A tireless campaigner for an end to ecological madness, he purchased several hectares of karri forest in the southern tip of Western Australia to prevent it being cleared by loggers. He and his wife built a dwelling there using rammed earth, and their only sources of power are sunlight and a wood-burning stove.

The Blairs also have a more conventional family home in Perth, where they often accommodate visiting scientists who have come to pick David's brains. Blair has devoted his working life to honing the techniques for superaccurate measurements of various sorts. His principal goal is to detect the collision of black holes and neutron stars in the depths of space. This daunting task can be achieved, he believes, by measuring the passage of gravitational waves. Predicted by Einstein in 1916, these elusive ripples have yet to be detected on Earth. Theoretically, however, the clamor of stellar collisions should reverberate through the universe in the form of gravitational waves. By measuring the almost inconceivably small sound vibrations produced when such a wave passes through the lab, Blair hopes to be able to tune in to an event that would otherwise go completely undetected.

As a by-product of this ambitious project, Blair and his colleagues have constructed the most accurate clock in the known universe. To be fair, it holds that record for only about five minutes at a time; atomic clocks and pulsars are more stable over longer intervals. The Blair clock is based on an ultra-pure crystal of sapphire, shaped like a fat spindle a few centimeters across. If the sapphire is struck, it will ring with the purest tone, and therein lies the secret of the clock: the crystal oscillations play the role of a pendulum, marking out intervals of time with unrivaled precision. The trick is to couple the crystal to an oscillating electronic circuit, using feedback to maintain frequency stability to the highest possible fidelity. The electrical oscillations are then fed into a digital display so that the operator can tell the time.

To work reliably, the entire apparatus must be cooled to near absolute zero ($-273°$ Centigrade) using liquid helium. To keep cool, the clocks must be entombed in glorified thermos flasks—these are the blue cylinders I mentioned. The sapphire itself sits inside a cavity made of niobium metal, chosen because it becomes superconducting at low temperatures. This turns the cavity walls into highly efficient mirrors. Microwaves from the electronic circuitry are fed into the cavity, where they create a distinctive pattern of standing electromagnetic waves at a frequency designed to resonate exactly with the crystal oscillations. Of course, this basic design is augmented by many refinements—circuits to stabilize the temperature, extra metal containers, shielding to prevent leakage of microwaves, a device to recycle the helium as it evaporates, and so on. The end product is a clock that is accurate to better than one part in a hundred trillion over three hundred seconds. Now, this is so good that, if you put one sapphire clock at the top of the Empire State Building and another at the bottom and compare, you ought to be able to tell the difference. In practice, you would have to synchronize two clocks on the ground floor, take one up in an elevator, wait a while, return it to the ground, and compare the readout with the clock left behind. There should be a slight but measurable mismatch. It would be a beautiful and direct test of Einstein's theory. Blair has plans to do just this, but there are technical obstacles, such as how to avoid too much disturbance to the clock's workings resulting from all that movement.

An easier test is to take a clock much higher, where the temporal mismatch is greater. In 1976, Robert Vessot and Martin Levine of the Smithsonian Astrophysical Observatory used hydrogen maser clocks to measure the effect of gravity on time. They put one clock in the nose-cone of a Scout D rocket and launched it to a height of ninety-six hundred kilometers from Wallops Island in Virginia. By monitoring the progress of the rocket-borne clock by radio, and comparing with a similar device on the ground, they could demonstrate the changing timewarp. The experiment was complicated because warps came from both the movement of the rocket and the change in gravity, and the two effects had to be separated in the data analysis. On the way up, the motion effect was initially dominant, but as the rocket climbed higher the speed fell and the gravitational effect grew. Vessot tracked the spacecraft for two hours before the payload finally plunged into the sea off Bermuda, totally destroying the delicate apparatus. Because his maser clocks were accurate to one part in a million billion, whereas the gravitational time-warp is four parts in ten billion, the effect could be checked to very high accuracy. Vessot was able to confirm Einstein's 1907 prediction, seventy years later, to seventy parts in a million. Not only does time really run faster at higher altitudes, it does so at just the rate that Einstein always said it would.

THE ECHO THAT ARRIVED LATE

Originally Einstein suggested looking for the telltale signs of the gravitational red shift in sunlight. Time on the sun's surface runs about two parts in a million slower than on Earth, because of the sun's much higher gravity. The spectral lines of the sun should therefore be shifted in frequency by the same factor. In practice, the effect, which should be readily detectable, is muddled up with other processes and is not very convincing on its own. However, the sun's timewarp shows up very clearly in another way, where it can be accurately measured. As the Earth and planets go around the sun, the positions of the planets viewed from Earth change in the sky. Sometimes a planet appears to be very close to the sun (where it can't normally be seen). If the planet lies on the far side of the sun, a beam of light traveling from the planet to Earth must then pass close to the sun's surface, where time runs a bit slower. The light will therefore arrive on Earth a little bit later than it would have done had the sun been in some other part of the sky. The basic geometry is shown in Fig. 3.5.

Einstein himself was aware of this time delay already in 1911, but did not follow up on its consequences. This had to await the inspiration of a young American physicist, Irwin Shapiro, who in 1961 calculated that the solar time delay would be measurable using radar rather than light. Radar waves can be bounced off other planets and returned to Earth in the form of a faint echo—in the same manner used to detect aircraft, but on a grander scale. If a planet is suitably located in the sky near the sun, the radar beam makes a double pass close to the solar surface, and according to Einstein's theory its echo should be delayed by a few hundred microseconds.

In 1959, Shapiro had been involved with the first recorded radar echo from the planet Venus. However, it was not until 1964 that the possibility

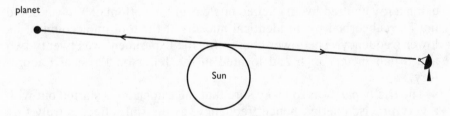

planet

Sun

3.5 When the sun almost interposes itself between a planet and Earth, radar signals and their echoes must travel close to the solar surface, where time runs slightly slower. This delays the echoes by a few hundred microseconds. Einstein's theory also predicts that the radar beam (or a light beam) will be slightly bent.

of measuring the relativistic time delay became feasible. In that year, the Haystack radar antenna in Westford, Massachusetts, became operational, with a power output of four hundred kilowatts. Shapiro published his idea later that year in the leading journal *Physical Review Letters,* and shortly afterwards he and his colleagues laid their plans. By late 1966, they were ready, and the Haystack antenna was turned on Venus, then Mercury, during favorable conjunctions. Hundreds of echoes were recorded and carefully analyzed by computer. Shapiro and colleagues found that the echoes were indeed delayed, and they were able to confirm the prediction of Einstein's theory to within 20 percent.

Errors in the radar-ranging experiment come from the rough topography of the target planet's surface, from uncertainties in the positions of the planets, and from distortions produced by the sun's corona. The technique can be considerably improved by using a spacecraft to return the signal. In 1977, NASA landed two Viking spacecraft on the surface of Mars, and this enabled much more accurate ranging data to be obtained. By 1978, Shapiro was able to confirm Einstein's time delay to one part in a thousand.

GOING UP IN THE WORLD

Remarkably, the first accurate test of the effect of gravitation on time was actually carried out on Earth, entirely within the confines of Harvard University! In 1959, Robert Pound and Glen Rebka decided to measure the gravitational red shift using not visible light but gamma rays. Like light and radio waves, gamma rays are electromagnetic waves, and can also serve as a clock. In this case, the frequency is millions of times greater, promising higher accuracy. The Pound-Rebka setup is reminiscent of the *perpetuum mobile* conveyor, but here atomic nuclei are excited rather than the electrons in the atoms, and there is no actual conveyor belt. The experiment was limited to demonstrating that a gamma ray emitted by an excited nucleus at the bottom of a tower could not be reabsorbed by an identical nucleus at the top, on account of its lower frequency. The tower chosen for the experiment was twenty-two and a half meters high and located in the Jefferson Physical Laboratory.

The tricky part was to make sure that the gamma rays started out with a very precise energy, hence frequency, as the shift effect is truly tiny (a mere two and a half parts in a thousand trillion). Normally when a nucleus emits a gamma-ray photon, the energy of the photon varies a lot, because the nucleus recoils and grabs some of the energy. To circumvent this complication, Pound and Rebka used a crystal of radioactive iron. In a crystal, the atom of interest is locked firmly into the lattice, so the ejected photon is able to brace itself against the whole mass of the crystal. The recoil is therefore negligible, and the gamma ray sets out with a very sharply defined frequency.

The experiment was readily able to show that, on moving to the top of the tower, the photon was too weak to excite any iron nuclei in a similar crystal there. To clinch the matter, Pound and Rebka wiggled the crystal at the top of the tower up and down. In the reference frame of the crystal, the wiggling motion creates a varying Doppler shift in the up-coming photon. By carefully arranging for the Doppler blue shift caused by the crystal going down to cancel the gravitational red shift due to the photon going up, the experimenters could induce nuclear excitation. The speed of the wiggle that brought about this state of affairs then gave a value for the gravitational timewarp. Pound and Rebka were thereby able to confirm Einstein's prediction to 1 percent.

When Einstein first predicted that gravity would slow time, all these delicate experiments lay in the far future. He was not unduly disturbed that such an important prediction had scant hope of being tested experimentally. As I have already explained, Einstein believed more in the power of thought than the power of experimentation to help us unravel nature. In June 1907, with the special theory of relativity now firmly established, and his mind preoccupied with the nature of gravitation, Einstein began thinking about an academic career. Today it would be astonishing that someone with Einstein's track record at that time would not have a secure university appointment, but the wheels of academia turned slowly in the Europe of 1907.

As a first step in his ascent of the professional ladder, Einstein applied for the right to teach at the University of Bern. This was known as a privatdozentship, and carried no salary. As he was not independently wealthy, Einstein was obliged to remain in the employment of the Swiss Patent Office. The committee that considered his application decided to reject it, on account of a technicality: Einstein had submitted seventeen published papers, but had omitted a piece of unpublished work, required by the rules. It was not until January 1908 that the application was finally accepted.

Einstein's teaching career began on a low note. He gave a course of lectures on heat in the summer of 1908 to just three students, which included his friend Besso. By 1909, however, the scientific community was waking up to the fact that they had a genius in their midst, and Einstein secured the position of associate professor of theoretical physics at the University of Zurich. The faculty board was almost unanimous in its decision to create the position specially for Einstein, even though he was a Jew and strongly anti-Semitic comments had been made in connection with the appointment. In July of that year, Einstein resigned from the Patent Office. He took up his new job in October, having by that time received the first of many honorary doctorates, from the University of Geneva.

The next few years of Einstein's career were certainly productive, but no great advances were made on the nature of time, or gravitation. He continued refining his thoughts on these subjects, and working towards a great synthesis, but that was still some time away. Meanwhile, he was busy teaching and attending meetings. Also, in 1910, Mileva gave birth to their second son. In 1911, the Einstein family moved to Prague, where Einstein obtained the post of full professor. Later that year, he began reformulating his arguments about the effects of gravitation on light and time, and wrote a series of papers. Einstein clearly recognized that he needed to generalize his special theory of relativity to take into account gravitational fields and accelerated motion, but he did not yet know how to do it. He confessed to Besso that he found the task "devilishly difficult."

A breakthrough came in mid-1912, about the time the Einsteins moved back to Zurich, where Albert took the post of professor at his alma mater, the ETH. Einstein came to the conclusion that a fully satisfactory general theory of relativity could be obtained only by giving up the normal rules of geometry. It was wrong to think that gravitation *causes* a distortion or warping of time, he realized—gravitation *was* a warping of time! More generally, *both* space and time must be warped. A gravitational field is not a field of force at all, but a curvature in the geometry of spacetime.

Einstein knew next to nothing about curved geometry, but he had a mathematician friend, Marcel Grossmann, who taught Einstein the necessary techniques, which had been worked out by Gauss and Riemann in the nineteenth century. All the elements for a general theory of relativity were now available, but they remained frustratingly scattered in Einstein's mind. Slowly and painfully over a year or two, Einstein and Grossmann inched towards the final synthesis.

During this period, moves were afoot to persuade Einstein to move to Berlin. Several job possibilities were discussed, and in late 1913 Einstein accepted formal membership of the Prussian Academy of Sciences. The Einsteins left Zurich in March 1914, as Europe lurched towards war. It was to be a decisive break with the past. Einstein was to remain in Germany until 1932, but Mileva soon announced that she and their two sons were returning to neutral Switzerland. The marriage had never been a particularly happy one, and divorce followed. Biographers suggest that Albert was already at this stage having an affair with his cousin Elsa, whom he later married. Anyway, he returned for a while to a bachelor existence and pronounced himself never more happy.

In contrast to the dismal collapse of his marriage, Einstein's scientific work was approaching a sensational climax. He was still struggling to piece together space, time, matter, motion and gravitation into a consistent mathematical scheme, but within a few months he would succeed, and the world would know the stunning implications of his general theory of relativity.

CHAPTER 4
BLACK HOLES: GATEWAYS TO THE END OF TIME

There exist in the heavens therefore dark bodies, as large as and perhaps as numerous as the stars themselves.

PIERRE DE LAPLACE (1796)

WARP FACTOR INFINITY

The hands of the clock have stayed still at half past eleven for fifty years. It is always opening time in the Sailors Arms.

DYLAN THOMAS, *Under Milk Wood*

I once received a letter from a man in Thailand asking, in all seriousness, whether paradise could be reached through a black hole. Since the evocative term was coined by Princeton physicist John Wheeler in 1967, black holes have assumed an almost mystical appeal for the public. Perhaps it is their ability to suck in and imprison anything that approaches them that provides the fascination.

The black hole is the ultimate test of Einstein's ideas. Although the existence of gravitational timewarps has now been thoroughly confirmed by delicate experiments on Earth and in the solar system, the effects are incredibly tiny and of little practical relevance outside of navigation and astronautics. If they were the only consequences of the general theory of relativity, this aspect of Einstein's work would largely be ignored today. As it happens, there exist many objects in the universe that warp time in spectacular fashion.

In 1967, a young Englishwoman named Jocelyn Bell accidentally stumbled on a gravity timewarp a million times greater than that produced by the sun. She accomplished this feat using little more than chicken wire. As a graduate student of the Cambridge University radio astronomer Anthony Hewish, Bell was constrained by a tight budget for her research. She and Hewish wanted to study twinkling radio sources, and rather than use an expensive radio telescope they made their own by stringing chicken wire hopefully across a green Cambridgeshire field, in the proudest traditions of British science. One day Bell was perplexed by a fuzzy trace on the pen-and-ink output from the "detector." She noticed it occurred each night around midnight. Bell alerted Hewish, and together they studied the phenomenon more carefully. They soon came to the conclusion that the fuzzy traces were created by a radio source in space emitting regular pulses. Jocelyn Bell had discovered pulsars.

As I mentioned in Chapter 2, a pulsar is produced by a spinning neutron star, an object so compact that its gravitational field is a billion times stronger than Earth's. The effect on time is dramatic. Time at the surface of a typical neutron star is slowed by about 20 percent relative to Earth time. It is an arresting thought that, from the point of view of an observer on the surface of a neutron star (maybe not as daft an idea as it sounds—see Chapter 13), the Earth is only about 3.5 billion (Earth) years old, and the universe is two or three billion years younger than we estimate.

Time is warped so drastically on a neutron star because the star, while possessing the mass of the sun or more, is nevertheless compressed to a radius of just a few kilometers. The stronger the gravity at the surface of an object, the more time is slowed, or stretched. In Fig. 4.1, I have graphed the timewarp factor against radius for an object with the mass of the sun. The distinctive feature is the way that the curve shoots off the top of the graph as the radius approaches about three kilometers. This is a critical radius. The timewarp does not simply get very large when the body is compressed to this radius; it actually becomes infinite.

It is amazing that the essential idea of this infinite timewarp was known to Einstein almost as soon as he had formulated his general theory of relativity. Yet it was not his own discovery. As I explained in the last chapter, Einstein struggled for years attempting to generalize his theory of relativity to include the effects of gravitation on space and time. Many mathematical descriptions were tried, only to be discarded as unsuitable. In the autumn of 1915, while Europe was embroiled in war, the pacifist Einstein was embroiled in advanced mathematics. After a final intensive burst of study, he at last hit upon the system of equations that ensured his immortality. Today they are known as the Einstein gravitational-field equations.

On 2 November, in triumphant mood, Einstein addressed the Prussian

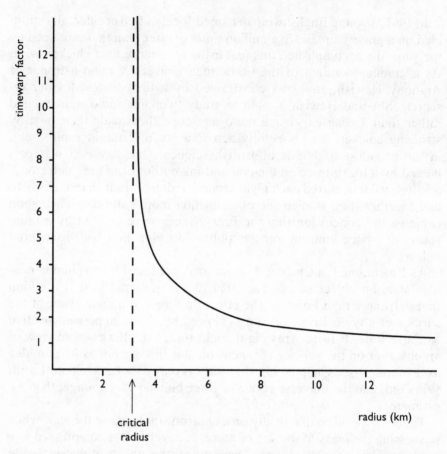

4.1 Escalating timewarp. The graph shows the warp factor at the surface of a spherical object of one solar mass as a function of radius. The timewarp rises towards infinity as the critical radius of about 3 km is approached.

Academy of Sciences in Berlin and presented the equations that today bear his name. By 16 January 1916, he was back, reading a paper by one Karl Schwarzschild, director of the Potsdam Observatory. Schwarzschild himself was unavailable. He was actually busy fighting on the Russian front, and was soon to die of an illness contracted there. The paper was a landmark, in that it contained the first exact solution to Einstein's brand-new field equations. Schwarzschild had been following Einstein's work on gravitation and was able to read about the final form of the field equations in the 25 November issue of the *Proceedings of the Prussian Academy of Sciences*. He was immediately ready with a solution describing a simple but physically useful example: the gravitational field in the empty region outside a uniform ball of matter.

Schwarzschild's solution was just what was needed to calculate the gravitational field in the vicinity of the Earth and sun, for which it yielded a fully convincing account. Significantly, it correctly reduced to Newton's law of gravity at large distances. However, there was something odd about the solution that baffled Einstein all his life. Indeed, another forty years was to elapse before the significance of this oddity was fully appreciated. To understand the problem, it is first necessary to have some idea of what Einstein's equations are designed to do. First, they relate the strength of the gravitational field at each point in space to the matter and energy that creates the gravity. A solution of the equations assumes a particular distribution of matter and energy (in Schwarzschild's case, a ball of matter) and gives the corresponding gravitational field produced by it. But there is more. Because a gravitational field is associated with a warping of time, a particular solution also tells us how much time is dilated at each point in space. In the case of Schwarzschild's famous first solution, the timewarp is given by a very simple formula that depends only on the distance from the center of mass. It is this formula that I have used in drawing Fig. 4.1.

The runaway timewarp, located near three kilometers in the case of the sun, was the thing that troubled Einstein. The existence of a critical radius, now called the "Schwarzschild radius," at which time is *infinitely* dilated, seemed to him deeply unphysical. In terms of the behavior of light, any photon attempting to climb away from the critical radius would be infinitely red-shifted—its frequency and energy reduced to zero, effectively rendering it impotent. An observer far from the star would not be able to see anything at all. However hot the star glowed at its surface, from a distance it would appear black.

Einstein was well aware in 1916 that Schwarzschild's solution contained this puzzling feature, but at the time it scarcely seemed a serious problem. The radius of the sun is seven hundred thousand kilometers—nearly half a million times larger than its Schwarzschild radius—and Schwarzschild's solution does not apply to the region *inside* the sun. If the curve is relevant to the external region only, it must be truncated far to the right in Fig. 4.1, where the timewarp has climbed to only very modest proportions (about two parts in a million). To warp time seriously, the sun would have to be compressed to a size much smaller than Earth. Such a prospect would have seemed utterly fantastic to the Prussian Academy of the day.

A DARK MYSTERY

I think there should be a law of Nature to prevent a star from behaving in this absurd way!

ARTHUR EDDINGTON

The conservatism of Einstein and his associates had not been shared by some adventurous earlier scientists. True, the Earth and the sun are both very much larger than their Schwarzschild radii, but what about other astronomical bodies? Might they be either massive enough or compact enough to create an infinite timewarp? Incredibly, as long ago as 1784, an obscure English clergyman by the name of John Michell suggested essentially just that. In a paper communicated that November to the Royal Society of London, he wrote:

> If there should really exist in nature any bodies whose density is not less than that of the sun, and whose diameters are more than 500 times the diameter of the sun . . . their light could not arrive at us.[1]

Michell knew nothing, of course, about timewarps or the general theory of relativity. He based his calculation on Newton's theory of optics, which assumed that light consists of particles, or corpuscles, and on his theory of gravitation. However, it so happens that Newton's theory concurs with Einstein's in certain conclusions about the effect of gravity on light. "Let us suppose the particles of light to be attracted in the same manner as all other bodies with which we are acquainted," Michell speculated. He then reasoned that, for a particle to escape permanently from a gravitating body, it had to be projected from the surface of that body with a certain minimum speed—its so-called escape velocity. In the case of the Earth, for example, the escape velocity is 11.2 kilometers per second. If a body is propelled away from the Earth at a slower speed than this, it will eventually fall back down.

The escape velocity from a spherical body depends on both its radius and its mass. If the Earth retained all its material but were compressed to one-quarter of its size, the escape velocity would be doubled; you would need to project an object at least as fast as 22.4 kilometers per second for it to fly off into space for good. Similarly, if the Earth had more mass, the escape velocity would be higher. Michell noted that if the mass of a body of given radius were large enough the escape velocity would exceed the velocity of light. Under these circumstances, light would not be able to flow away, and the body would appear black. The formula that Michell wrote down for this state of affairs is, remarkably, identical to the formula connecting the mass and the Schwarzschild radius.

Michell's conclusion was echoed a few years later by the famous French mathematician Pierre Laplace. In spite of Laplace's standing as a scientist and scholar, nobody took the speculation on "black stars" too seriously for quite a while. By a curious coincidence, in the very month when Karl Schwarzschild was writing down his famous solution, the astronomer Walter Adams, working at the Mount Wilson Observatory in California, announced that he had obtained a spectrogram of the light from a puzzling star named Sirius B. Sirius is the brightest star in the sky, but it has a very faint companion whose existence was deduced in 1834 from the way that Sirius appears to be disturbed by the gravitational field of an unseen object nearby. Adams was astonished to find that the spectrum of Sirius B differed hardly at all from that of Sirius. This could mean only one thing: Sirius B was as hot as Sirius. Why, then, was it so dim? The answer had to be that it is very small—about the size of the Earth. A hot star as small as Earth implied a density many thousands of times higher than that of ordinary matter.

The idea of a star squashed to such a tiny size was greeted with consternation when Adams suggested it. However, over the years it gradually dawned on astronomers that extremely compressed stars were not only possible but inevitable. But black stars? In 1921, Sir Oliver Lodge, a distinguished British scientist well known for his psychical research, lectured to students at the prestigious Birmingham University, where he had been the first principal. He told the students that, "if light is subject to gravity, if in any real sense light has weight . . . a sufficiently massive and concentrated body would be able to retain light and prevent its escaping."[2] He then went on to compute some numbers: "If a mass like that of the sun could be concentrated into a globe of about 3 kilometers in radius, such a globe would have the properties above referred to; but a concentration to that extent is beyond the range of rational attention." Nevertheless, Sir Oliver noted that a massive galaxy confined to a few hundred light-years in diameter would trap light even though its average density would be only a thousand-trillionth of that of water. "This really does not seem an utterly impossible concentration of matter."

By the time Sir Oliver delivered his address, Einstein's fame had spread across the world. In 1919, the highly respected British astronomer Sir Arthur Eddington had managed to confirm a key prediction of the general theory of relativity—the very slight bending of a starbeam by the sun's gravity—and the public appreciated the irony that a British scientist had confirmed a "German" theory not long after the ending of wartime hostilities. The word "relativity" soon entered the popular vocabulary. Einstein was fêted and showered with honors. Invitations flooded into Berlin from many countries. Everyone wanted to meet him, from millionaires and show-biz personalities to politicians. Academically, his career was approaching its zenith. In 1921, he received Fellowship of the Royal Society of London—a rare honor for a foreigner. The following year, while en route to Japan, Einstein was awarded the Nobel Prize. He gave the money to Mileva, as part of their divorce arrangements. The prize was bestowed, however, not for relativity, but for Einstein's 1905 work on the photoelectric effect.

Einstein also visited the United States, where he met President Harding at the White House. He was besieged by newspaper reporters and hailed as a hero. The name Einstein became synonymous with genius. Actually, his trip to America was not made for scientific reasons, but was primarily a political exercise. Although Einstein had never taken his Jewishness very seriously and remained a sort of reverent atheist throughout his life, he was befriended by Chaim Weizmann and persuaded to take an interest in Zionism. The visit to America was to raise funds for the Hebrew University in Jerusalem.

In spite of the fame Einstein had by then attracted for himself and his theory of relativity, there were many opponents. Partly this was because the predictions of the theory remained extremely hard to test. However, there was a lot of hatred for Einstein because of his race, his politics (a mixture of Zionism and pacifism) and his nationality. Others were jealous of his accomplishments. Many simply didn't understand his work. He once remarked, cynically: "If my theory of relativity is proven successful, Germany will claim me as a German and France will declare that I am a citizen of the world. Should my theory prove untrue, France will say that I am a German, and Germany will declare that I am a Jew."[3]

Although the general theory of relativity gradually gained acceptance in the 1920s, it had a reputation for being impenetrably difficult. It was said that only Einstein, Eddington and one or two others fully understood it. This was undoubtedly an exaggeration, but the paucity of measurable applications worked against the theory, and it became something of a specialist scientific backwater. In fact, it was not until the 1930s that its consequences began to be taken more seriously. By this stage, astronomers had come to accept the existence of extremely dense stars like Sirius B, dubbing them "white dwarfs," but there was violent disagreement about what might happen to a star that was more massive or more compact than a white dwarf. Would its intense gravitational field cause it to shrink still further? Could it approach its Schwarzschild radius, and if so what would prevent it from collapsing completely, under its own weight, to a point of infinite density?

In 1930, a bright nineteen-year-old Indian student, Subramanyan Chandrasekhar, was toying with the equations describing a white-dwarf star during a long sea voyage to England, where he intended to work at Cambridge University with the great Eddington. To his bafflement, Chandrasekhar found his sums predicting an outlandish result. If the white dwarf had a mass of more than 1.4 suns, then, according to the calculation, it could not remain stable, and would collapse further, without any apparent limit. On arrival in England, Chandrasekhar showed his calculation to British astronomers, who dismissed it as an unimportant oddity. A typical white-dwarf star, in spite of its compaction, is still thousands of times larger than its Schwarzschild radius. For many, the idea that a body with the mass of the sun could become compressed into a ball only a few kilometers across just did not appear credible. In fact, it was downright repugnant.

Werner Israel, an international expert on black holes, has carried out a historical study of the attitudes of scientists to gravitational collapse, with its attendant infinite timewarp. He finds deep-seated psychological and philosophical prejudices at work, creating resistance to the idea even today:

> As the web of observation and theory slowly tightened, the scientific reaction—first disregard, then dismay, yielding only gradually to the beginning of acceptance—set a pattern for the disclosures yet to come.[4]

Einstein's own attitude remained intransigent. As late as 1939, he wrote that the infinite timewarp "does not appear in nature for the reason that matter cannot be concentrated arbitrarily."[5] How much Einstein and his peers were subconsciously influenced in their skepticism by atavistic belief in the absolute solidity and indestructibility of atoms—a belief that stretches back to Greek times—is unclear. Already from the time of Schwarzschild's solution, it was known that a body with a radius less than $1\frac{1}{8}$ times its Schwarzschild radius cannot possibly resist collapse, even if it is made of supposedly incompressible material, because the central pressure would become infinite. Yet the very thought of unrestrained total implosion was too disturbing for most scientists to contemplate. The ever-voluble and opinionated Eddington was unequivocal in his expression of distaste. The Schwarzschild radius he described as "a magic circle which no measurement can bring us inside."[6]

PENETRATING THE MAGIC CIRCLE

Behold! The jaws of darkness do devour it up.

WILLIAM SHAKESPEARE

Einstein's blunt dismissal of the possibility that a star might retreat within its Schwarzschild radius was to prove a fateful statement. Just two months later, Robert Oppenheimer, later to be Einstein's boss at the Institute for Advanced Study in Princeton, submitted a paper to *Physical Review,* co-authored with his student Hartland Snyder, entitled "On Continued Gravitational Attraction." The contents boldly addressed the problem of what would happen to a massive star once its nuclear fuel finally ran out. The star would then be unable to sustain a high enough internal pressure to withstand its enormous weight. The paper opens with the prophetic statement: "When all thermonuclear sources of energy are exhausted, a sufficiently heavy star will collapse."[7] The ensuing calculations, based on Einstein's gravitational-field equations, led the authors to conclude that the collapse would "continue indefinitely," and the star would plunge on through the critical radius, creating an infinite timewarp on the way.

Oppenheimer's seminal work on collapsing stars was largely overlooked as the world descended into another war. The subject of nuclear energy became more than just a branch of theoretical astrophysics, and Oppenheimer himself was recruited to run the atomic-bomb project. There was still a widespread feeling that the Schwarzschild radius implied a physical impossibility. It was, after all, mathematically singular —that is, it described a physical quantity becoming infinite. There is an unwritten rule in science that when anything potentially observable is predicted to become infinite it is a sure sign that the theory is breaking down somehow. Einstein himself interpreted this infinity to mean that the particles constituting the star would be obliged to move faster than light if the star were compressed so much, and he ruled this out on account of his special theory of relativity. Before the infinite timewarp set in, something would have to give.

After the war came the cold war, and the scientific community was divided into East and West. According to Israel:

> In Western circles the work of Oppenheimer and Snyder was a forgotten skeleton in the cupboard by the 1950s, and the notion of gravitational collapse, if anyone had even thought of raising it, would have been dismissed as the wildest speculation; but in the Soviet Union it was an orthodox textbook result.[8]

Israel cites the 1951 Soviet edition of Landau and Lifshitz's *Statistical Physics,* a classic student textbook, for a description of the fate of a star that has exceeded the maximum mass at which it can resist its own weight:

> It is clear from the start that such a body must tend to contract indefinitely. . . . From the point of view of a "local" observer the substance "collapses" with a velocity approaching that of light, and it reaches the center in a finite proper time.[9]

In the West, it was largely John Wheeler at Princeton University, a freewheeling and imaginative physicist who had worked with Niels Bohr and alongside Einstein, who put the subject of gravitational collapse on the scientific map. But there remained the stubborn enigma of the Schwarzschild radius and the "magic circle" that for so long had suggested an absolute no-go zone, protected by a physical barrier of some unknown nature.

With the benefit of hindsight, we can now see that the puzzle of the infinite timewarp has been tacitly solved many times over in the course of history, before it eventually dawned on a bewildered scientific community. Already in 1916, a Dutch physicist named Johannes Droste, who independently discovered Schwarzschild's solution, recognized that the critical radius had to be taken seriously. Eddington unwittingly solved the problem in the 1920s, and Oppenheimer and Snyder gave a clear enough exposition in their famous 1939 paper. But the message of all these contributions was ignored amid the general prejudice, championed by Einstein himself, that untrammeled implosion through the Schwarzschild radius was physically impossible.

A SINGULAR PROBLEM

Go on till you come to the end; then stop.

Lewis Carroll

Mathematicians often confront infinity in their equations, and rarely bat an eyelid. They call infinite points in their solutions *singularities*. It is significant that it needed mathematicians rather than physicists to solve the riddle of the critical radius. This did not happen until about 1960, when the situation was at last clarified, independently, by Martin Kruskal and David Finkelstein in the United States, and George Szekeres in Australia. All three recognized that the singular nature of the Schwarzschild radius was purely a mathematical artifact: nothing *physically* singular occurs there at all. None of the scientists involved was active in mainstream relativity theory. Finkelstein had his own agenda concerning novel relationships between mathematics and physics and was working largely in isolation. Kruskal was a young researcher at Princeton who performed a back-of-the-envelope calculation on the geometry of Schwarzschild's spacetime, largely for amusement, after a seminar on the subject. He showed the result to Wheeler, somewhat diffidently, thinking it too trivial to deserve publication. It was left to Wheeler to publicize it. Szekeres began his career as a chemical engineer in Hungary, from where he fled the Nazis in the late 1930s to settle in

Shanghai. Left alone by the Japanese during the occupation, he ended the war working for the Americans as a clerk, and took to doing mathematics, his real love, during his spare time. After a few years, he obtained a lecturing post in mathematics at The University of Adelaide. He became interested in general relativity largely because it was a convenient application of certain mathematical techniques he was developing, and he too did not regard his solution of the Schwarzschild singularity problem as especially significant. As a result, he published it in an obscure Hungarian journal, where it lay almost unnoticed for some years.

What these mathematicians all hit upon is that the mathematical singularity at the critical Schwarzschild radius is analogous to what happens to the lines of latitude and longitude at the north and south poles on a map of the globe. The standard Mercator projection depicts Antarctica and Greenland as severely distorted; distances become stretched more and more as the poles are approached. In reality, of course, the geometry is no different at the poles from anywhere else on the Earth's surface. The illusion of distortion arises purely because of the latitude-and-longitude coordinate system employed. Schwarzschild's coordinates suffer from the same problem as Mercator's. By the simple expedient of transforming to a new set of coordinates, one can make the mathematical singularity at the Schwarzschild radius go away, and it is possible to penetrate (mathematically!) Eddington's "magic circle."

I don't understand, complains our patient skeptic. Surely there is an infinite timewarp at the Schwarzschild radius? That sounds both pretty physical and pretty singular to me.

The crucial thing is that no *local* physical quantity is singular at the Schwarzschild radius. A timewarp involves a *nonlocal comparison* of clock rates: you have to compare clocks at the Schwarzschild radius with clocks far away to discover it. If you were actually at the Schwarzschild radius, you wouldn't say: "Oh! Time is infinitely warped here." In fact, you wouldn't notice anything odd at all about time—or any aspect of local physics—in your immediate vicinity. It is only by comparing your time with someone's elsewhere that you discover the timewarp.

To clarify this point, let us resurrect our intrepid twins Ann and Betty. Suppose Ann stays on Earth and Betty goes by spaceship to the vicinity of the imploded star, armed with a clock. Imagine that the star has shrunk inside its Schwarzschild radius and compressed itself to a tiny ball; we shall shortly see what its ultimate fate might be. Fig. 4.1 tells us the rate of passage of Betty's clock relative to Ann's in the case for which the collapsed mass is equivalent to that of the sun. From the graph you will learn, for example, that, when Betty is six kilometers from the center of mass, her clock runs at half the speed of Ann's. They can check this by sending radio signals to each other. To avoid mixing the timewarp due to motion with that for gravitation, we can imagine that Betty uses her ship's powerful rockets to remain motionless at a fixed distance from the collapsed star. (A real human being could not withstand the tremendous g forces this would entail.) Betty then finds that she ages less rapidly than Ann, and Ann agrees. There is no symmetry about their circumstances in this scenario; it is definitely *Betty* who is subjected to a large gravitational field and its associated stretching of time. Ann and Betty can compare clocks, dates and experiences to convince themselves that for Betty time really is "running slowly" compared with Ann. If Betty speaks, Ann will hear her words stretched into a low-pitched drawl. She will see Betty's clock running at half-speed. All other physical processes will appear in slow motion too, Betty's speed of thought and aging included.

Betty herself notices nothing unusual about her speech, her mind, her aging or the passage of time. Everything in her immediate vicinity looks normal. However, when she uses her telescope to observe the Earth, events there seem to be racing along at twice the normal pace. If Betty observes Ann's clock, it will seem to her to be running at two hours to her one. Ann's words will be high-pitched and compressed, like those old Chipmunk songs. Physical processes around Ann will appear to Betty to be taking place in double-time, as if she were viewing a video recording stuck on fast-forward. All of this is real, not some bizarre optical illusion. Betty can fly back home and directly compare ages and clocks with Ann. The results will confirm that, for Betty, time really has been slowed by her trip to a region of intense gravity. She will have aged by half the amount that Ann has.

Perhaps this is so. It is, after all, just an amplification of the effect that Vessot and others have measured on Earth. But what if Betty ventures close to the Schwarzschild radius itself? Surely something odd would happen? The timewarp is supposed to be *infinite* there. How can anything physical be truly infinite?

As Betty hovers closer to the Schwarzschild radius, so the timewarp grows ever larger. She will see Ann's clock racing faster and faster ahead of her own. Ann will see Betty's clock increasingly slowed. Betty will feel even more uncomfortable, because to hover closer to the massive object she must endure truly enormous g forces. Something else starts to become important too. The same slowing of time that affects Betty's clocks also affects the light waves emitted by atoms of the spaceship, and the radio waves that Betty uses to talk to Ann. These waves suffer a large gravitational red shift, which means that the ship starts to look very red to Ann. In fact, it looks positively dim, even if Betty takes the trouble to floodlight it for Ann's benefit; the hard climb up through the ferocious gravitational field saps the light of so much energy that only a feeble image survives. Also, Ann has to tune her radio to a very low frequency to pick up Betty's slurred speech.

The closer Betty gets to the Schwarzschild radius, the more severe these effects become. But, although in principle Ann from afar would see events in the spaceship running very slowly, in practice she has a hard time seeing anything at all, because of the spiraling red shift. This rises without limit as Betty approaches the Schwarzschild radius. The intensity of the light correspondingly diminishes towards zero. Betty and her ship fade from sight completely. As John Michell predicted two centuries ago, when Ann looks in the direction of the collapsed star, all she sees is blackness: a black hole.

Why is it called a black *hole* exactly?

The ability of Betty to hover close to the Schwarzschild radius depends on her rocket's having the power to resist the colossal pull of gravity there. The g forces assaulting her and the spaceship rise without limit the closer the ship hovers to the critical radius; in effect, Betty's weight becomes infinite there. In this sense, Einstein was right: there would be something unphysical—locally—about this *static* state of affairs. It would certainly be physically impossible for the ship to hover at exactly the Schwarzschild radius, where time grinds to a halt relative to Ann's frame of reference. In reality, no force in the universe is strong enough to withstand the pull of gravity there. Well before the limiting radius was reached, the rocket motors would inevitably lose their battle against gravity, and the ship would plunge across the Schwarzschild radius. Once the ship is inside, there is no longer *any* possibility of its being able to hover at a fixed distance from the center of gravity. The same applies to any matter there, including the material of the imploded star: it will all fall to the center. The region inside the Schwarzschild radius cannot contain any static matter. So this region of spacetime is generally *empty* and appears *black* (from the outside): hence the term "black *hole*."

But surely a powerful enough motor could resist any gravitational force, however large?

Not in this case. Gravity is so fierce inside the black hole, it even plucks back *outward*-directed light—just as Michell conjectured—pulling it inescapably towards the center of gravity as if focused by a gigantic lens. Amazing that this image was captured already in 1920 by a certain A. Anderson of University College Galway, who wrote in the *Philosophical Magazine* of that year:

. . . if the mass of the sun were concentrated in a sphere of radius 1.47 kilometres, the index of refraction would become infinitely great, and we should have a very powerful condensing lens, too powerful indeed, for the light emitted by the sun itself would have no velocity at its surface. Thus . . . it is shrouded in darkness.[10]

For the ship (or the material of the imploded star) to resist this astonishing gravitational force and remain at a fixed radius would imply that it was traveling faster than light. This is forbidden by the theory of relativity.

Hang on. The ship wouldn't be moving at all, relative to the center of gravity—or to Ann. Why do you say it would be traveling faster than light?

The concept of speed is essentially a local one. For example, you can measure the relative speed of A *passing* B, but if A and B are spatially separated and in a gravitational field, their relative speed becomes a rather vague thing. If you try to measure it by sending light signals, timing the motion between points in space and so on, you run into the problem of whose clock to use. And when A is on Earth and B is inside a black hole, you can't make any sense at all of the speed of B relative to A. Granted, police officers measure speed-at-a-distance with their radar traps, but this is because they can neglect the effects of gravitation. If they tried measuring *vertical* speed, they would run into trouble. In that case the Doppler shift would get all mixed up with the gravitational red shift, and determining a well-defined speed would become a problem. (Not much of one in the case of the Earth, of course, because the gravitational red-shift effect is so small, but you can see my point.) By contrast, you *can* make sense of the relative speed between a spaceship and a light pulse passing close by it, even when both are inside the black hole. This speed must always be the speed of light. So, if the *outward*-directed light is nevertheless being dragged *inwards*, the ship must be too, or it would be *locally* traveling faster than light, which is against the special theory of relativity.

If the gravity of the collapsed matter pulls on light, doesn't that mean outgoing light from just outside the Schwarzschild radius travels out towards Ann at a reduced speed?

Yes and no. Certainly light from the region close to the black hole can take a very long time to get out as measured by Ann. Remember, however, that she cannot directly measure the speed of light at a distance. She can deduce that the photons are plodding their weary way towards her with painful slowness by the standards of *her* clock, but when she takes into account that time near the Schwarzschild radius is slowed, she finds that, in the region where a given photon happens to be, it is moving at that same old speed—three hundred thousand kilometers per second. And, of course, an observer near the black hole would be able to measure the speed of light locally and get the same answer.

Doesn't this mean that Ann will never see Betty actually reach the Schwarzschild radius and fall into the black hole?

That's right. The light conveying the position of Betty's ship takes longer and longer to reach Ann as the ship gets nearer and nearer to the Schwarzschild radius. You can see from Fig. 4.2, which compares Betty's clock with Ann's clock as Betty's ship free-falls into the black

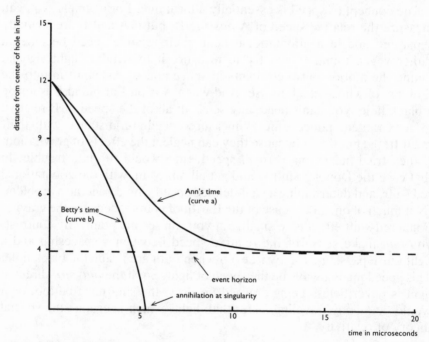

4.2 Falling into a black hole. Graphed here are clock readings taken by Betty as she falls into a solar-mass black hole (curve b), and Ann as she watches from afar (curve a). Both curves are drawn as a function of Betty's distance from the center of the hole. Betty starts out at rest at a distance of 12 km. She reaches the Schwarzschild radius (event horizon), represented by the broken line, after 4.7 microseconds, and hits the central singularity after 5.3 microseconds. By contrast, Ann sees Betty's fall drastically slowed, and become "frozen" forever just outside the Schwarzschild radius. The region below the broken line is there-fore "beyond the end of time" as far as Ann is concerned.

hole, that Ann would have to wait an infinite length of time to see Betty slip across the fateful radius.

So, while Betty perceives herself to be plunging madly into the hole, Ann will see Betty's ship just hovering there—frozen in space and time, so to speak?

Not really. Because of the escalating red shift and dwindling light intensity coming up from the ship as it nears the Schwarzschild radius, Ann cannot follow the onset of this "freezing." What she actually sees is the ship getting rapidly redder and dimmer, until it simply merges into the inky blackness of the black hole. In effect, it is swallowed up by the blackness.

But it still means that, in reality, Betty *never ever* reaches the Schwarzschild radius? The black hole is a bit of a fiction, isn't it?

Not so! Seductive words like "never" and "ever" have no absolute meaning. You have to specify whose "never" and whose "ever" you are talking about. True, in Ann's frame of reference Betty never reaches the Schwarzschild radius, but in Betty's frame of reference she certainly does. In fact, it would typically take Betty just a few microseconds, spaceship time, to fall into a solar-mass black hole from a distance of a few times the Schwarzschild radius. She makes it all right, and in pretty quick time too!

But black holes are still a fiction as far as we in the universe outside are concerned. The star that implodes to supposedly create the black hole would also take an infinite amount of our time to retreat inside its Schwarzschild radius. The black, supposedly empty, region would *actually* be occupied by the remnants of the star, wouldn't it?

In a sense that is correct. In fact, the Russians originally disliked the term "black hole" for that very reason. Out of earshot, they would jokingly call them "labor camps," since nothing can come out of them. Officially they used the term "frozen stars," acknowledging that from a distance time and motion are frozen at the Schwarzschild radius. Now, you might be able to *deduce* that, technically, what looks like a black hole is "in reality" a star frozen in the very late stages of collapse. But all the properties of this collapsing star become very rapidly (typically in milliseconds or less from the onset of collapse) indistinguishable from a *genuinely* empty, already-formed black hole. And if you ventured close to check out any residual difference, you would find none, because you would end up following the imploding material of the star across the Schwarzschild radius into real emptiness. So, you see, the distinction between a black hole and a frozen star is practically a vacuous one.

BEYOND THE END OF TIME

Einstein's equation says "this is the end" and physics says "there is no end."

JOHN WHEELER

Our skeptic speaks:

Let's imagine Betty being inside the black hole for a moment. If it takes *forever* (Ann's ever) for Betty to cross the Schwarzschild radius and enter the hole in the first place, *when* (Ann's when) is it that Betty can be inside the hole? Isn't Betty somehow beyond the end of time (Ann's time) there?

That's about it. When you have an infinite timewarp, one woman's microsecond is another woman's forever. The interior of the black hole is a region of space and time that can never be observed from the outside. The Schwarzschild radius separates events inside the black hole, which are forever unobservable by Ann however long she waits, from events outside the black hole, which Ann can witness if she is patient enough. The Schwarzschild radius is often called the "event horizon" for this reason. So, in a certain sense, the interior of the black hole is beyond the end of time as far as the outside universe is concerned—though not beyond the end of Betty's time, of course.

This means that, in the few microseconds that Betty takes to whiz across the horizon, all of eternity will have passed by outside. The fast-forwarded images will accelerate to *infinite* speed. Betty will know, once inside the hole, that the universe outside is "over," even if it has lasted *forever!*

Just as with the earlier twins experiment, it is best not to think in terms of "what is Ann doing now" and so on. If Betty looks up at the universe outside, she will not actually see an infinite duration passing in that fleeting microsecond of her time, because the light takes time to reach the black hole from distant cosmic events, and before most of it gets there she will have plummeted to the center of the hole, and oblivion. The only way Betty could witness the infinite future history of the universe would be to hover at the Schwarzschild radius and wait for all the incoming light to arrive. But we have seen it is impossible to hover at the Schwarzschild radius.

If Betty is beyond the end of time, what happens to her? Is it possible for her to turn around and come back out of the black hole?

It goes without saying that you can't come back from beyond the end of time without going backwards in time. This is probably not on (though see Chapter 10). Betty seems to have two possible fates. The safest bet is that she falls straight to the center of the black hole and is annihilated. If Schwarzschild's solution is followed doggedly to the exact geometrical center, it predicts that the gravitational field becomes infinite there—the center of the black hole is a spacetime singularity. Unlike at the critical radius, this central singularity can't be conjured out of existence by a change of coordinates. It is local and physical—rather than merely mathematical—in character. If the singularity exists, then it will be a boundary to time itself, an edge of infinity where time ceases to exist and there is no beyond. In which case Betty's plunge into the black hole will be a one-way journey to nowhere—and nowhen. When she smacks into the singularity, she cannot continue in spacetime, so she must cease to exist as a physical entity. She will, of course, already have been crushed to oblivion by the time she gets there.

A wilder idea is that the black-hole interior might be more complicated in a way that enables Betty to miss the singularity and survive. Alternatively, some yet-to-be-discovered physics may prevent the singularity from forming. In either case Betty will obviously continue in spacetime, but she cannot go into any region in "our" space, because "its" time has already passed. The only possibility is for Betty to emerge in some *other* space—another universe, if you like—that is connected to ours through the interior of the black hole. This other space would be a universe located beyond the end of time as far as we are concerned. To lapse poetic, it is, perhaps, the closest science has come to identifying a candidate for the Land Beyond Time, and perhaps explains why the man from Thailand asked the question about paradise. Unfortunately, there is no reason to believe (a) that such a region of spacetime actually exists, (b) that you could really "go through a black hole" to get to it even if it did exist, and (c) it would turn out to be very different from our universe if you did. There is also the problem that, if you could do this "tunneling" trick once, you could do it again—in the other universe. But as you can't fall from the other universe back into ours (without traveling back in time), it would be necessary for you to discover a third universe, then a fourth, and so on. You have to assume the potential existence of an *infinity* of almost-disconnected universes, an idea which delights some people but strikes me as plain absurd.

"Absurd" just about sums up the whole subject. I think Eddington was right. Surely the whole idea of black holes and infinite timewarps is just too crazy to take seriously? Where is the evidence that these things actually exist?

ARE THEY REALLY OUT THERE?

The light is shifted to the red. It becomes dimmer millisecond by millisecond, and in less than a second is too dark to see. . . .

JOHN WHEELER

In the early 1970s, I was a young lecturer at King's College in London. The local expert on black holes was John G. Taylor, an amiable mathematical physicist and professional actor with research interests ranging from the brain to supergravity theory. It was largely from the success of John's popular book *Black Holes: The End of the Universe?*, coupled with his good looks and easygoing media style, that the British public first became aware of black holes—"the most awesome objects known to man," as Taylor liked to remark. Unfortunately, the publication of the book coincided with a surge in interest in matters paranormal in the British media. The weird properties of black holes had a certain mystical appeal, and some people seemed to regard these objects in the same light as black magic. The aura of mystery was reinforced by certain conservative scientists who denounced the whole notion of black holes as speculative bunkum.

In spite of this, evidence was already accumulating that black holes had to be taken more seriously than magic. It was always clear that there is no *theoretical* impediment to a black hole forming: given a big enough mass, the Schwarzschild radius is so large that a black hole will form before the matter is compressed to an unusual density. The relevant question is, will the necessary conditions come about in the real universe?

Astronomers focused their attention mainly on dead stars, taking their cue from Oppenheimer's work on stellar collapse. The basic scenario is clear enough: When a star runs out of fuel, it will shrink under its own weight. If the star is heavy enough, no force can prevent it imploding into a black hole—according to Einstein's general theory of relativity. Back-of-the-envelope calculations suggest that stars with masses greater than about three suns will inevitably suffer this fate, assuming they do not find a way to blow off some material first. Many stars are known with three or more solar masses, so the existence of black holes as stellar remnants seems, on the face of it, entirely reasonable.

By 1960, astronomers had a good idea of how catastrophic stellar collapse might come about. A heavy star gobbles up its fuel at a prodigious rate, and when the supply runs out it doesn't just shrink. Instead, the core of the star implodes suddenly. The resulting shock releases a pulse of energy huge enough to blast the outer layers of the star into space. A titanic explosion ensues. Such outbursts have been observed by astronomers throughout history; they are called "supernovae." This seemed a good place to start in the hunt for gravitationally collapsed objects.

However, at that time scientists still had a mental block about the nature of the Schwarzschild radius. Kip Thorne, then a student of John Wheeler, explains:

Perhaps nothing was more influential in preventing physicists, between 1939 and 1958, from understanding the implosion of a star than the name they used for the critical circumference: "Schwarzschild singularity." The word "singularity" conjured up an image of a region where gravity becomes infinitely strong, causing the laws of physics to break down—an image that we now understand is correct for the object at the center of a black hole, but not for the critical circumference.[11]

There was a widespread feeling that everything somehow ground to a halt at the critical radius, that the infinite timewarp that freezes events to a standstill from the viewpoint of a distant observer also spelled the end of the road for the collapsing star in some vague and unspecified sense. It needed the calculations of Finkelstein, Kruskal and Szekeres to convince scientists finally that in the reference frame of the in-falling matter nothing actually stopped at the Schwarzschild singularity. As Thorne points out, "a person who rides through the Schwarzschild singularity (the critical circumference) on an imploding star will feel no infinite gravity and see no breakdown of physical law."

Notwithstanding his famous taste for the wild and bizarre, Wheeler himself was initially skeptical about total gravitational collapse. It was around 1960, with the personal knowledge of Kruskal's "removal" of the Schwarzschild singularity, and impressed by the latest ideas about supernovae, that he changed his mind. Thorne recalls how, one day in the early 1960s, Wheeler rushed into a relativity class late, beaming with pleasure. He had just returned from a visit to the Livermore Laboratory in California, where Stirling Colegate, the world's top supernova expert, had unveiled the latest computer simulations. Colegate had based his computations on Oppenheimer and Snyder's classic prewar calculation, but had included many more realistic features. "With excitement in his voice, he drew diagram after diagram on the blackboard," writes Thorne. Wheeler explained how the imploding core of a medium-mass star would produce a neutron star, but for a core heavier than about two solar masses it seemed that nothing could arrest the collapse:

> As seen from the outside, the implosion slowed and became frozen at the critical circumference, but as seen by someone on the star's surface, the implosion did not freeze at all. The star's surface shrank right through the critical circumference and on inward, without hesitation.[12]

This realization that the critical radius would not prevent total collapse was the turning point. But the concept of the black hole, not to mention the name, was not yet fully developed. The 1960s proved to be a decade of great ferment in astronomy, much of it related to the muddled subject of gravitational collapse. First came the discovery of quasars, or quasi-stellar objects (QSOs). These pinpricks of light, located in the far reaches of the cosmos, were initially mistaken for stars. By 1963, they were recognized to be as massive as galaxies and enormously bright, but also incredibly compact. Their discovery forced astronomers to confront the possibility that such massive, dense objects risked undergoing total gravitational collapse.

The following year, a high-altitude rocket equipped with a primitive X-ray detector registered a strong X-ray source in the constellation of Cygnus, the swan. It was dubbed Cygnus X-1, and ten years later it became the first candidate object for a possible black hole formed by stellar collapse.

The early and mid-1960s also marked important theoretical advances. British mathematician Roger Penrose developed new and much slicker geometrical techniques for studying Schwarzschild's spacetime, event horizons, collapsing stars, singularities and related aspects of the general theory of relativity. These new methods were to prove a boon for physicists struggling to come to terms with the bewildering properties of black holes.

Finally came the discovery of pulsars (neutron stars) in 1967. By this stage gravitational collapse, supernova implosions, frozen stars and infinite timewarps were firmly on the astrophysicists' agenda. In late 1967, a conference on pulsars was held in New York, and Wheeler referred to the possibility that continued collapse would produce a "black hole" in space. The name had finally entered the English language. But that was only the beginning. More important was finding irrefutable evidence that black holes really exist in the universe.

In the early 1970s, as black-hole research blossomed into a world-wide industry, astronomers began seeking them in earnest. The use of satellite-borne X-ray telescopes greatly improved the understanding of objects like Cygnus X-1, and suggested that if a black hole forms in a binary-star system it will betray its presence by slowly eating its companion, shining brightly with X-rays as a result. At the time of writing, many astronomers regard Cygnus X-1 as very probably a black hole trapped in a tight 5.6-day orbit around a blue supergiant star.

Quasars and disturbed galaxies have proved another promising place to search for black holes, but in this case the objects concerned would be considerably more massive than a collapsed star. Indeed, astronomers suspect that the cores of some galaxies may harbor black holes with masses equivalent to millions or even billions of suns. There is good evidence that at least one million-solar-mass black hole lurks at the center of our own Milky Way. Though a single really convincing candidate remains frustratingly elusive, the accumulated evidence for black holes has become overwhelming in the last few years. Eight decades after Schwarzschild found his famous solution, the existence of real objects with infinite timewarps seems at last to be confirmed.

Einstein never lived to see the fruitful applications of his general theory of relativity to collapsed stars. All the evidence suggests he was, in any case, wary of the entire subject. In fact, Einstein largely lost interest in the "local" effects of gravitation around 1920, following Eddington's successful test of the bending of light by the sun. The Roaring Twenties saw the birth of quantum mechanics—a challenging field that engaged Einstein's full attention. Meanwhile, the new hundred-inch telescope at Mount Wilson in California became available for systematically studying the most distant objects in the universe. Over the years, the astronomers involved increasingly noticed something odd about the light from these objects. By the end of that decade, it was clear that Einstein's general theory of relativity had found a new and even more dramatic application: to the origin and evolution of the universe itself.

CHAPTER 5
THE BEGINNING OF TIME: WHEN EXACTLY WAS IT?

The beginning of time fell on the beginning of the night
which preceded the 23rd day of October, in the year 4004 B.C.

BISHOP JAMES USSHER (1611)

The beginning of the movement [was] on Sunday, at sunrise,
1,974,346,290 Persian years having passed up to the present.

PETRO D'ABANO

THE GREAT CLOCK IN THE SKY

Not far from the millionaires' playground of St.-Tropez in the affluent south of France lies a rambling stone château set in magnificent wooded grounds. A short walk through the trees from the main house stands a collection of modern cottages which serve as accommodation for guests. The château itself opens onto a huge stone veranda overlooking well-tended gardens, complete with three swimming pools and a circular stone-walled arbor for outside entertaining. The interior of the château is tastefully appointed in traditional style. Potted plants, priceless paintings and a huge antique grand piano adorn the massive lounge. Adjoining the main building, a small lecture room offers the latest in audiovisual aids. The house is called Les Treilles, and it is owned by the family of Anne Schlumberger, an elegant lady with a discerning taste for cultural artifacts and fine wine.

In the summer of 1988, Madame Schlumberger hosted an unusual gathering at Les Treilles. The guests were scientists, and all had, in one way or another, an interest in the subject of time. Some of them were mavericks who had decided to reject the conventional wisdom about time, the universe and (almost) everything. Most were physicists or astronomers: Geoffrey Burbidge from California, Vittorio Canuto from New York, David Finkelstein from Atlanta. The meeting was convened by the Belgian chemist Ilya Prigogine, whose own unusual ideas about time had both thrilled the general public and exasperated many fellow scientists.

This was the summer when Stephen Hawking had also just shot to international fame, with the publication of his book *A Brief History of Time*. It had already climbed effortlessly into the best-seller lists worldwide, and was destined to remain there, in Britain at least, for five years —an all-time record for any book. Most of the participants of the Les Treilles meeting were convinced that the subject of time would have a much longer history than Hawking supposed.

Hawking's brief history of time is really a brief history of the universe, based on the assumption that time began when the universe began. But the title of Hawking's book also implies something else: that the universe *has* a meaningful history. A coherent account of "what happened to the universe" supposes that we can discuss the cosmos as a whole and talk about it changing overall, stage by stage, from what it was to what it is. Can we do that?

Einstein thoroughly mixed things up with his discovery that there is no universal time, no master clock that monitors the heartbeat of the cosmos. Time is relative: it depends on motion, it depends on gravity. But the universe is full of both. The Earth goes around the sun at thirty kilometers per second, the sun orbits the galaxy at 220 kilometers per second, the galaxy moves in the local group of galaxies at a similar speed. More important, the galactic clusters themselves fly apart, caught up in the general expansion of the universe, so that the most distant galaxies appear to be moving away from us at nearly the speed of light. In addition to this ubiquitous motion, all astronomical bodies possess gravitational fields, some of them immense and drastically timewarping. Given the existence of a myriad times, how can we talk about the universe as a whole marching through history to the beat of a single cosmic drum?

In a higgledy-piggledy universe full of chaotic motion and haphazard concentrations of matter there would indeed be no well-defined cosmic history, for there would be no universal time. Fortunately and mysteriously, on the largest scale of size the universe is not chaotic. Both the distribution of galaxies and their pattern of motion, when averaged out, are surprisingly uniform. A good marker of this uniformity is provided by the background heat radiation that fills space. Discovered by Arno

Penzias and Robert Wilson in 1965, this microwave radiation pervades the universe, and is widely believed to be the afterglow of the hot big bang in which the universe originated. The year after the Les Treilles gathering, NASA launched a satellite called COBE (for Cosmic Background Explorer) to study this thermal bath. The COBE scientists found that the background radiation is uniform across the sky to an accuracy of one part in a hundred thousand. Because the universe is almost completely transparent to electromagnetic waves, the cosmic background radiation will have propagated undisturbed through space for billions of years. It is therefore a living relic, a leftover, from the primeval inferno that accompanied the birth of the cosmos. When we detect this radiation, we are observing the universe as it was about three hundred thousand years after the big bang. Any major irregularities in the universe will, because of the gravitational red-shift effect, leave their imprints in this radiation. Because the COBE data revealed no marked variations in the strength of the heat radiation from different regions of the universe, we may infer that the universe is, and almost always has been, extremely smooth on a large scale.

The COBE results also tell us something very important about Einstein's time. In fact, it isn't quite true that the cosmic background heat radiation is completely uniform across the sky. It is very slightly hotter (i.e., more intense) in the direction of the constellation of Leo than at right angles to it. There is a very good reason for this. Imagine traveling in a spaceship at high speed towards Leo. The radiation coming from that direction of the sky will be blue-shifted by the Doppler effect, whereas the radiation coming from the opposite side of the sky will be red-shifted. These shifts make the background radiation more intense in the Leo direction. In practice, Earth is our spaceship, whizzing through space, or, more accurately, whizzing through the enveloping bath of primeval heat, at about 350 kilometers per second. This makes the radiation appear wonky across the sky. But if you subtract out this so-called dipole anisotropy, then the resulting distribution is smooth to about one part in a hundred thousand.

Although the view from Earth is of a slightly skewed cosmic heat bath, there must exist a motion, a frame of reference, which would make the bath appear *exactly* the same in every direction. It would in fact seem perfectly uniform from an imaginary spaceship traveling at 350 kilometers per second in a direction away from Leo (towards Pisces, as it happens). This special state of affairs, this carefully selected view of the cosmos, singles out the reference frame of the imaginary spaceship as having a unique status. The time told by the spaceship's clock will also have a unique and special status. We can use this special clock to define a *cosmic* time, a time by which to measure historical change in the universe. Fortunately, the Earth is moving at only 350 kilometers per second relative to this hypothetical special clock. This is about 0.1 percent of the speed of light, and the time-dilation factor is only about one part in a million. Thus, to an excellent approximation, Earth's historical time coincides with cosmic time, so we can recount the history of the universe contemporaneously with the history of the Earth, in spite of the relativity of time.

Similar hypothetical clocks could be located everywhere in the universe, in each case in a reference frame where the cosmic background heat radiation looks uniform. Notice I say "hypothetical"; we can imagine the clocks out there, and legions of sentient beings dutifully inspecting them. This set of imaginary observers will agree on a common time scale and a common set of dates for major events in the universe, even though they are moving relative to each other as a result of the general expansion of the universe. They could cross-check dates and events by sending each other data by radio; everything would be consistent. So cosmic time as measured by this special set of observers constitutes a type of universal time, rather as Newton originally assumed was true for *all* observers. It is the existence of this pervasive time scale that enables cosmologists to put dates to events in cosmic history—indeed, to talk meaningfully at all about "the universe" as a single system.

THE BIG BANG, AND WHAT HAPPENED BEFORE IT

Who cares about half a second after the big bang; what about the half a second before?

FAY WELDON

The year 1924 saw Einstein hard at work in Berlin, happily married to his cousin Elsa. His scientific interests had shifted away from time and gravitation to quantum physics, which was about to occupy the center stage of science for the next decade or so. But in that same year, a discovery was quietly taking place in America that would turn out to have the most profound implications for Einstein's time. Mount Wilson in California is home to the hundred-inch Hooker telescope. Completed in 1918, it was the largest telescope of the day, and the only one capable of settling a long-running dispute concerning the structure of the universe. The wrangle centered on the nature of those peculiar fuzzy patches of light known as nebulae.

Since antiquity, astronomers had been intrigued by the milky splodges in the sky. In addition to the great sweeping arc of the Milky Way itself, there are three smaller patches of light visible to the unaided eye: the Andromeda nebula, and the two Magellanic clouds, called Large and

Small. Telescopes of even moderate size reveal hosts of these nebulae, and they attracted the attention of many astronomers over the years. Nobody knew what they were, but in the nineteenth century a French astronomer by the name of Charles Messier painstakingly compiled a catalogue of them, largely for the purpose of distinguishing nebulae from comets, which were considered to be much more interesting. Brighter nebulae carry the label "M" in Messier's honor. Andromeda is known as "M31."

As late as the 1920s, astronomers failed to agree on what the nebulae were. Two theories were current. One held that the Milky Way galaxy, made up of billions of stars including our sun, was the principal cosmic system. According to this view, all the nebulae were either gas clouds or distant clusters of stars that lay within the Milky Way galaxy or just beyond. The opposing theory was that at least some of the nebulae are huge star systems in their own right, similar to the Milky Way, and located at immense distances.

In early 1924, a young American astronomer named Edwin Hubble decided to settle the matter. Tall and somewhat imperious, Hubble had started life as a lawyer, but turned to astronomy, where he was to reveal a cosmic law that would be hailed as the discovery of the century. Using the great telescope at Mount Wilson, Hubble patiently scrutinized the nebulae M31 and M33. The power of the hundred-inch instrument was enough to resolve the images of individual stars in these nebulae. Hubble was soon able to find a distinctive type of variable star familiar to astronomers from our own galaxy. This star gave him a way to estimate the distance to M31, the Andromeda nebula. The answer came out to be about a million light-years. There could no longer be any doubt: Andromeda lay well beyond the edge of the Milky Way galaxy, and was clearly an entirely separate galaxy in its own right, comparable in size and shape to our own. Hubble went on to identify other familiar stars in Andromeda. Astronomers rapidly accepted that the universe was vastly larger than they had previously thought, with other galaxies scattered through space as far as telescopic observations would permit.

For many years prior to Hubble's work, astronomers had tried to illuminate the controversy by photographing the light spectra of nebulae. The leading expert at that time was Vesto Slipher, a dedicated assistant of Percival Lowell, the astronomer who had founded an observatory at Flagstaff, Arizona, to search for the canals of Mars. Lowell, believing the nebulae were solar systems in the process of forming, set Slipher the job of confirming this spectroscopically. One of the useful things a spectrum will do, thanks to the Doppler effect, is to reveal information about the motion of the source. In 1912, Slipher determined that M31 is moving *towards* the Earth at three hundred kilometers per second. By 1917, he had obtained spectroscopic velocity data on twenty-five nebulae having distinctive spiral shapes (like our Milky Way). All but four displayed *red* shifts indicating that, unlike Andromeda, they are rushing away from us.

The preponderant outward movement suggested that some sort of systematic effect was at work, but Slipher had no means to determine the distances of his nebulae to demonstrate this. Furthermore, the prevailing belief concerning the organization of the universe was that it formed a static system, with the Milky Way located at its center and the nebulae subordinated to it. With Hubble's discovery, however, the mood began to change: it was now possible to measure the distances to the galaxies. Hubble himself set about obtaining distance and red-shift data for a few dozen galaxies. It gradually became clear that the more distant galaxies systematically displayed larger red shifts, indicating that they are moving away from us faster. By 1929, Hubble was able to announce one of the most momentous scientific discoveries of all time: the universe is expanding.

Hubble based his sensational claim on the red-shift data, indicating that the speed with which a galaxy recedes from us is directly proportional to its distance. This means galaxies twice as far away are moving at twice the speed. Hubble's "law" is only statistically correct, as individual galaxies can have quite large random velocities around this overall "Hubble flow" (recall that Andromeda is actually moving towards us). But suitably averaged over many galaxies there is an unmistakable mathematical relationship between speed and distance. The particular proportionality law discovered by Hubble can be interpreted to mean that the galaxies are moving away from *each other* as well as away from the Milky Way. In other words, the entire assemblage of galaxies is dispersing. This is what is meant by saying the universe is expanding. Since Hubble's original observations, it has been found that galaxies tend to cluster in groups which do not expand, and may even contract. Nevertheless, on a scale of clusters of galaxies and larger, the universe is definitely expanding. In addition, the pattern of expansion is highly uniform: on average, it is the same in all directions. It is this uniformity that is reflected in the smoothness of the cosmic background heat radiation.

Clearly, if the universe is growing bigger, it must have been smaller in the past. We can imagine running the great cosmic movie backwards until all the galaxies are squashed together. This compressed state corresponds to the time of the big bang, and in a certain sense the expansion of the universe can be considered as a vestige of that initial explosion. Today it is normal for cosmologists to claim that the universe *began* with the big bang. This weighty conclusion follows if you trace the expansion back in time to some idealized point of origin at which all the matter of the universe is concentrated in one place. Such a state of infinite density represents an infinite gravitational field and infinite spacetime curvature

—i.e., a singularity. The big-bang singularity is similar to the situation at
the center of a black hole that I described in the previous chapter, but
lying in the past rather than the future. As it is not possible to extend
space and time through such a singularity, it follows that the big bang
must be the origin of *time itself*.

People, especially journalists who get angry about scientists explaining
everything, often ask: What happened before the big bang? If this theory
is correct, the answer is simple: *nothing*. If time itself began with the big
bang, there was no "before" for anything to happen in. Although the
concept of time being abruptly "switched on" at some singular first
event is a hard one to grasp, it is by no means new. Already in the fifth
century, Augustine proclaimed that: "The world was made, not in time,
but simultaneously with time."[1] Keen to counter jibes about what God
was doing before he made the universe, Augustine placed God outside
of time altogether, making him the creator of time itself. As I described
in Chapter 1, the idea of time coming into being with the universe therefore fits very naturally into Christian theology. In Chapter 7, we shall
see that recent ideas in quantum physics have changed our picture of the
origin of time somewhat, but the essential conclusion remains the same:
time did not exist before the big bang.

OLDER THAN THE UNIVERSE?

When Hubble presented his data in 1929, nobody drew any sweeping
conclusions. The evocative term "big bang" did not come into use until
much later. Astronomers nervously shied away from discussing the ultimate origin of the universe, and were content merely to acknowledge
that the highly compressed earlier state must have been very different
from what we see today. Nevertheless, although the physical significance of the big bang was unclear at that time, Hubble's data did enable
scientists to work out a rough date for this event by measuring the *rate*
at which the universe is expanding. If Hubble's data are taken at face
value, the date works out to be 1.8 billion years ago. More precisely, if
the universe has always expanded at the same rate as today, Hubble's
observations suggested that the galaxies would have been all compressed
together 1.8 billion years ago.

However, before we jump to conclusions, we first have to ask whether
the rate of cosmological expansion has changed over time. The universe
does not expand freely: the galaxies attract each other with gravitational
forces, and this will operate to restrain their dispersal, thereby slowing
the rate of expansion. Fig. 5.1 shows the general manner in which the
expanding universe slows with time as a result of this gravitational braking effect. Graphed here is the size of a typical region of space as a

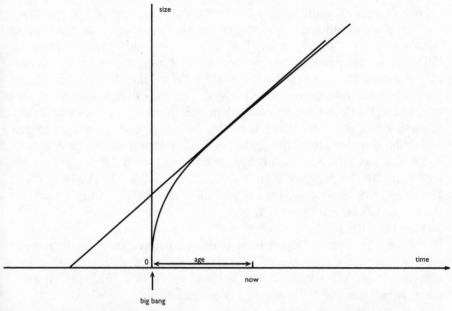

5.1 *The graph shows how the universe changes its rate of expansion with time. It begins by expanding very rapidly just after the big bang, but progressively decelerates as a result of gravitational restraint. The slope of the graph at the present epoch (marked "now") is shown as a straight line, and represents the observed rate of expansion today. If this rate had remained constant over time, the projected age of the universe, obtained by extending the sloping line back to the time axis, would be considerably greater than the true age.*

function of time. The universe starts out with zero size and an infinite rate of expansion: this is the big bang. The steepness of the curve here indicates a rapid increase in size near the beginning. The graph then curves over steadily, tracing a gradual reduction in the rate of expansion as the size of the universe grows. The deceleration effect also reduces with time. This feature is easy to understand: gravity weakens with distance, so as the galaxies get farther apart the braking force diminishes.

The present epoch is marked on the graph. The slope of the curve at this point corresponds to the rate at which the universe is expanding today, the value of which is given by Hubble's law. That slope is marked on the figure by the straight line tangent to the curve. Had there been no braking effect, the straight line would be the correct history of the universe. Clearly, for a given present rate of expansion, a universe which brakes must be considerably younger than one which expands at a fixed rate, because when the line is extended back in time it reaches zero considerably to the left of the curve. Therefore, using Hubble's value of 1.8 billion years, we conclude that the universe would have to be considerably younger than this.

Hubble himself didn't press the point about the age of the universe. At that time, cosmology was scarcely a proper science, and profound conclusions were discouraged. But by the 1950s, this upper limit of 1.8 billion years had begun to worry people. The problem was simple. Radioactive dating had given a figure of 4.5 billion years for the age of the Earth. Putting this number together with Hubble's data led to the absurd conclusion that the Earth is older than the universe! Radioactive dating of meteorites (and more recently the moon) has given similar ages to that of Earth, compounding the embarrassing mismatch. But worse was to come. Over the decades, astronomers have built up a detailed picture of how stars age by burning their nuclear fuel. Among the oldest stars in our galaxy are those contained in globular clusters, and from a study of these it has been estimated that some stars have been around for at least fourteen or fifteen billion years.

The glaring discrepancy in time scales burgeoned from a nagging unease into a serious problem for cosmologists as they sought to inject some rigor into the subject in the postwar years. In a 1952 survey, Hermann Bondi had the following to say on the matter:

> The great importance of the time-scale difficulty for many of the cosmological theories has been stressed. The difficulty arises because the reciprocal of Hubble's constant, as deduced from the velocity-distance relation, is an appreciably shorter time than the ages of the Earth, the stars and meteorites, as determined by various methods. Owing to the crucial importance of this discrepancy for so many theories and models there is probably no other investigation of such significance for cosmology as further research on these time-scales.[2]

At the time, Bondi was backing the so-called steady-state theory of the universe, which he had helped fashion with colleagues Thomas Gold and Fred Hoyle in the late 1940s. This was a head-on attempt to circumvent the time-scale problem—by doing away with the origin of the universe altogether. In the steady-state model, the universe has neither beginning nor end. It goes on expanding forever, and as the gaps between the galaxies swell, more galaxies form to fill them, made from new matter that is continually being created out of nothing by some unknown process. The details are arranged to make the universe look more or less the same on a large scale at all epochs: there is no evolution, no changing cosmic history. Although the steady-state theory attracted numerous adherents for a while, when Arno Penzias and Robert Wilson discovered the cosmic background heat radiation in 1965, it knocked the stuffing out of the theory for good. The cosmic background is so neatly explained as a relic of the hot big bang, it is hard to imagine the universe has always existed in its present form.

Hermann Bondi was, incidentally, the first scientist I ever saw in the flesh. The occasion was about 1960, and he came to my high school in North London to give a special lecture on the theory of relativity and its implications for the nature of time. I can still remember his vivid description of how to synchronize distant clocks using light signals, delivered in his characteristic measured tones, enriched by the trace of a mid-European accent (he was originally from Vienna) that, since Einstein established the stereotype, somehow seems to lend credibility to scientific pronouncements.

EINSTEIN'S GREATEST MISTAKE

The biggest blunder of my life . . .

ALBERT EINSTEIN

Einstein himself lost track of cosmology in the 1920s, and seems to have learned of the expansion of the universe only after visiting Hubble in California in 1931. By this stage of his career, Einstein had become distracted with quantum mechanics, and was increasingly involved in international politics. With the rise of Nazism, the situation in Germany was deteriorating. As a Jew, a pacifist, and an independent thinker of international renown, Einstein was especially vulnerable. He sought more and more opportunities to travel abroad, making regular visits to Oxford University and the California Institute of Technology in Pasadena. It was on one of these visits that he met Hubble.

In the early days of relativity theory, Einstein had been keenly interested in cosmology. Following his formulation of the general theory of relativity in 1915, he soon produced a model for the large-scale structure of the universe, using his description of gravitation in terms of spacetime curvature. This was published in 1917. Nobody then suspected that the universe is expanding, so it was perfectly natural for Einstein to seek a model that was static and eternal. No matter that the stars would burn out after a few billion years; these were the early days of astrophysical theory, and physicists still had little idea of how the stars shine. The chief obstacle confronting Einstein in his early cosmological investigations concerned the very nature of gravity itself. As in Newton's theory, the general theory of relativity describes gravitation as a universal attraction, acting between all bodies in the cosmos. This leads to something of a paradox, because a collection of unsupported bodies all attracting each other cannot remain static: they will inevitably fall together into a single mass. In other words, the universe will collapse under its own weight.

To evade this grave difficulty, Einstein came up with an ingenious solution. He proposed that the force of gravitational attraction is opposed by a new type of repulsive force, fine-tuned in strength to exactly counterbalance the weight of the cosmos, thereby achieving a static equilibrium. Rather than simply put such a force into the theory by hand, Einstein examined his general theory of relativity for clues. The gravitational-field equations were not, of course, handed to Einstein on tablets of stone, nor were they somehow derived from Newton's theory. He arrived at these equations after years of nitpicking mathematics, taking into account many factors, including simplicity and elegance. The simplest versions of the field equations work admirably, correctly reducing to those of Newton when the gravitational fields are weak. They also lead to several successful predictions.

The principal shortcoming of Einstein's original field equations was that the gravitational force they describe is purely attractive, and is therefore inconsistent with a static universe. To circumvent this problem, Einstein made the fateful decision to add an extra term to his original field equations. He called it the "cosmological term." Although the cosmological term is simpler than the other terms in the equation, and is in certain respects a natural addendum, it represented in many eyes an ugly adulteration and had all the hallmarks of a fix. Worse still, the cosmological term enters the theory multiplied by an unknown number, called the "cosmological constant," usually denoted by the Greek letter Λ (pronounced "lambda"). The trouble about all this is that it is an unwritten rule in science to keep the number of independent quantities in the theory as small as possible. Newton's theory had just one undetermined constant, called "G," which is a measure of the strength of the force between two point masses. The numerical value of G is found by measuring the force of attraction between two heavy balls of known mass a measured distance apart. Einstein's theory also contains G, and now it had a second constant, Λ, also to be determined from measurement.

The cosmological term is optional in the sense that it can be removed simply by setting Λ equal to zero, thereby recovering the original field equations. But if Λ is chosen to be a positive number, the force it describes is repulsive, as Einstein desired. Being a component in an all-embracing theory of gravitation, the Λ force can be considered as a type of antigravity. The nature of the Λ force is, however, distinctly different from "normal" gravity, and other familiar forces. Most forces decrease in strength with distance, but the Λ force actually gets *stronger*. This has the virtue that the cosmological repulsion is negligible on the scale of the solar system, where Einstein's original theory already gives impressive accuracy, but still makes its presence felt over extragalactic distances.

A value for Λ can be worked out from the requirement that the repulsion is strong enough to counteract the weight of a given large region of the universe. From the known average density of cosmic matter Einstein was able to calculate how heavy a given region of the universe is, and thereby deduce Λ. It was easy to check that the cosmological term would be completely negligible in its local effects. For example, in the case of the Earth's gravity, the Λ force would reduce your weight by only a few billion-billion-billionths of a gram—less than the weight of a single atom. The Earth's attraction to the sun would be diminished by the equivalent of a gentle puff of air. So, although the Λ term might be considered by some to be artificial, *ad hoc* and ugly, it cannot be ruled out by appealing to local physics. The only way to test for it is by cosmological observation.

In the event, Λ turned out to fail in its intended purpose, for two reasons. First, it didn't do the job properly; second, it appeared to be unnecessary anyway. These shortcomings were exposed not by Einstein, who seemed to lose interest in cosmology just as it was becoming exciting, but by a number of European scientists. The most significant of these was a Belgian cleric and mathematician, Monsignor Georges Lemaître. Born in 1894, Lemaître worked all his life at the University of Louvain. Colleagues described him as a man of robust vigor with a stentorian laugh. He was decorated for valor in the First World War, and in the Second he exercised courageous leadership at the university during the German occupation, a service for which he was awarded Belgium's highest national honor. Although Lemaître made important contributions to celestial mechanics and the use of modern electronic computers for numerical analysis, he is best remembered as the man who turned the study of cosmology from a minor branch of physics into a respectable discipline in its own right. His theoretical investigations matched Hubble's work on the observational front and gave birth to the subject of scientific cosmology in a recognizably modern form.

Lemaître made full use of Einstein's gravitational-field equations in his investigations, but unlike Einstein he did not restrict himself to static solutions. In 1927, Lemaitre discovered that Einstein's proposed tug-of-war between gravitational attraction and the cosmological repulsion couldn't work, because it was unstable. The slightest disturbance would cause the universe either to collapse, or to embark on an unending career of runaway expansion, as either normal gravity or cosmic repulsion gained the upper hand. Perhaps more significantly, it was becoming increasingly clear by that time that the universe was in any case not static, but expanding.

When Einstein finally woke up to these facts, the effect was dramatic. He publicly recanted, abandoning his static model of the universe in utter disgust. Out too went the offensive and adulterating cosmological term which had been invoked especially to explain it. Einstein bemoaned the fact that, had Hubble made his discovery somewhat earlier, the cosmological term would never have been introduced. Indeed, had Ein-

stein stuck to the original equations and pursued the consequences fear-
lessly, he would surely have been led to predict the expansion of the
universe several years before it was actually discovered, which would
undoubtedly have been one of the greatest accomplishments in the history of science. As it happened, he was distracted by an overly conventional adherence to the notion of a static universe. It was a major missed
opportunity. Later Einstein was to describe the introduction of the cosmological term as the biggest blunder of his life.

With the benefit of hindsight, this reaction by Einstein can be seen as
emotional and hasty. True, the cosmological term was no longer required
for the purpose of explaining a static universe, but, logically, the fact
that the universe is expanding does not *rule out* a Λ force, it merely
makes it unnecessary for its original intended purpose. Einstein, in his
chagrin at having failed to predict the expansion of the universe, may
have thrown the baby out with the bathwater—as we shall see.

Lemaître showed that Einstein's field equations were consistent with
a variety of expanding cosmological models, most of which started out
with a big bang. Curiously, many of these models had already been
discovered in 1922 by a little-known Russian scientist, Aleksandr Friedmann. Born in 1888, Friedmann lived in St. Petersburg, and unlike Einstein he was an exceptional student, highly gifted in applied
mathematics. In 1913, he turned his talents to the subject of weather
forecasting, and went to work at the aerological observatory in Pavlovsk.
When war came in 1914, his meteorological expertise was put to work at
the front, where he also became a pilot. He went on to lecture on fluid
dynamics and weather prediction at Kiev, and thence to a job at the
Petrograd Geophysical Observatory. It was here that he became interested in the general theory of relativity as a sideline. He applied Einstein's field equations (including the Λ term) to the problem of a universe
filled uniformly with matter, and discovered that, in addition to Einstein's static solution, there was the possibility of expanding and contracting models too. He published his results in two papers, pointing out
that the static nature of Einstein's model was purely an assumption
unsupported by observations. Einstein's initial response was that Friedmann had simply made a mistake in his calculations. Later he published
a more considered reply in which he conceded that Friedmann had done
his sums correctly, and that his work was "clarifying." However, Einstein still dismissed the idea of a time-dependent universe, and Friedmann's prophetic work languished in obscurity for a decade.

Poor Georges Lemaître fared little better than Friedmann at first. Following a tour of the United States where he learned about Slipher's red-
shift measurements, he published a paper in 1927 containing results very
similar to Friedmann's, and in which he anticipated the Hubble law. He
tried to attract the attention of Einstein and others to his work, but the

unassuming priest was not taken seriously. It fell to Eddington to champion Lemaître's important contributions, some years later, when Hubble's results had transformed the subject.

To appreciate the significance of the work of Friedmann and Lemaître, it is necessary to know something about the relationship between the equations of a physical theory and their solutions. It often happens in science that a set of equations possesses many solutions, each describing a possible reality. To pick one, you have to decide which best fits the facts, or else appeal to some additional criterion, such as physical reasonableness or elegance. Friedmann and Lemaître started with Einstein's gravitational-field equations, assumed that the universe is uniformly filled with matter having certain simple properties, and produced a large set of solutions. Included were Einstein's original static model, and a variety of expanding and contracting models. Each solution represented a possible universe consistent with Einstein's general theory of relativity. The burning question was, which one best corresponded to reality?

Einstein himself didn't offer much help. Annoyed at his big blunder over the cosmological term, rattled by the growing acceptance of quantum mechanics among his colleagues, and worried by the threats of Nazism and the world slump, he had his mind on other things. In fact, he was about to quit Berlin, and Europe, for good. In America, a new Institute for Advanced Studies was founded in 1932, and Einstein, now in his fifties, was offered a post. Originally he agreed to divide his time between Princeton and Berlin. Just a year or two previously, he had arranged for a small family house to be built on a vacant plot of land in the village of Caputh, a short walk from the river Havel, where he liked to sail. The Einsteins were happy living in Germany, but the storm clouds were gathering; when he and Elsa set out for America in December 1932, he sensed they would never return. "Take a good look," he said to Elsa as they left the little house he loved. "You will never see it again."[3] And he was right. The following month, Hitler came to power in Germany, and Einstein was high on the list of undesirables. His house was searched for weapons, and he was to be frequently vilified by the regime. He promptly resigned from the Prussian Academy of Sciences and for the second time renounced his German citizenship (he retained Swiss citizenship). After a brief spell in Belgium, he set sail for the United States, and Princeton, which was to become his permanent home. Other than a brief trip to Bermuda for visa purposes, he never left American soil.

In spite of these tribulations, Einstein did discuss his preference for a particular Friedmann solution in a paper with the Dutch astronomer Willem de Sitter in 1932. The Einstein–de Sitter model remains the simplest of the Friedmann models without a cosmological term. However,

Einstein did not take a great deal of interest in mainstream cosmology after this, and it was left to Eddington, Lemaître, Hubble and others to confront the issue of the big bang and the tough question of the ultimate cosmic origin. The situation in the early 1930s was in any case confused, and suffered from a lack of communication between "nuts-and-bolts" astronomers on the one hand, and mathematical physicists well versed in relativity theory on the other.

In reminiscing about these early developments in cosmology, the British cosmologist William McCrea remarked: "I do not recall that there was a rush to try and fit a particular Friedmann-Lemaître model to the observations. The immediate interest was in the indication from Hubble's results that the universe had been in a very congested state apparently no more than about 2 billion years ago."[4] Regarding the age difficulty, McCrea remembers that both Hubble's estimate of the time of the big bang and the radioactive dating of the Earth were regarded as subject to revision. "What impressed astronomers and geologists was that they were of the same order. . . . Certainly, nobody at the time seemed to expect the models to reveal anything about the *creation* of the universe or its earliest moments."

TWO-TIMING THE COSMOS

We cannot catch the fleeting minute and put it alongside a later minute.

EDWARD MILNE

Quite another reason why the age problem did not ring alarm bells in those early years had to do with the very nature of time itself. We can compare the relative rates of two clocks by placing them side by side, or at least by signaling back and forth between their observers. But how can we compare today's rate of passage of time, so to speak, with what it was a billion or more years ago?

The problem here is this: how do we know that a superaccurate cesium-beam atomic clock, if left for a few million years, won't tick a little bit faster or slower than it does today? I don't just mean a *particular* atomic clock, but all atomic clocks. Even if we can contemplate the notion of a universal cosmic time, can we be sure that the great clock in the sky has been ticking away evenly from the beginning of time until today? If the cosmic clock itself were changing with time, this would completely compromise our estimates of the age of the universe. Einstein freed time from the shackles of Newtonian rigidity: We know that time can vary from place to place. Why not from time to time? Might this be a possible solution to the time-scale problem?

Injecting this new level of uncertainty muddies the waters considerably. If time is something measured by clocks, and if clocks vary *with* time, how can we ever know what time it *really* is? Such bewildering issues were subjected to a searching analysis by Edward Milne, who was the first incumbent of the Rouse Ball Chair of Mathematics at the University of Oxford, a position now held by Roger Penrose. Milne, described by his more charitable peers as a kind man of brilliant intellect, must be considered one of the pioneers of modern cosmology, but he chose to plow his own lonely furrow. He happily accepted Hubble's observations of the expanding universe, but rejected Einstein's general theory of relativity, preferring his own theory, which he termed "kinematical relativity." This annoyed a lot of people and attracted much criticism.

Central to Milne's whole approach was the conviction that the laws of physics ought to follow from the nature of the universe and not the other way about, as is conventional. He reasoned that, if you start with the way matter is distributed through the cosmos, the way the universe expands, and so on, then things like the laws of gravitation and electromagnetism ought to follow from these facts as logical deductions. If such a scheme worked, it would short-circuit the usual processes of science, such as experiment and observation, and enable the laws of the universe to be derived more or less by pure thought alone.

This was heady stuff, and might have been more appealing had it not become conceptually very technical. From Milne's mathematical morass there did indeed emerge, here and there, equations reminiscent of familiar physics. But there were also some peculiar conclusions. One of these concerned clocks and timekeeping. Milne divined from his study of the way clock rates might be compared at different places and times that there was no God-given cosmic time; in fact, there could be any number of different time scales (see the quote above). Based on his assumptions about the way the universe is put together, he deduced that there are two time scales of special significance, which is one more than normal. The idea was that certain physical processes operate according to one time scale while other processes are governed by the second time. Milne labeled the two times by the Latin and Greek letters t and τ. The former is allegedly the time told by atomic processes and light, so it applies to atomic clocks and the frequencies of light waves for example. On the other hand, τ time is supposed to apply to gravitational and large-scale mechanical processes such as the rotation of the Earth and its motion around the sun. The distinctive feature of the theory is that t and τ may start out the same but they gradually slide out of step. Fig. 5.2 shows how the two times are connected. For mathematically inclined readers, τ is the logarithm of t.

What does the multiplication of time scales mean in practice? Most significantly, it implies that atomic clocks gradually march ahead of

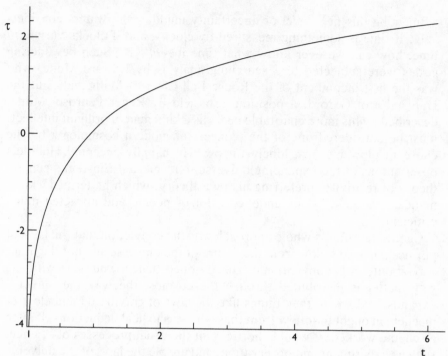

5.2 *Changing times. In Milne's theory of time, there are two separate time scales, t and τ. Clocks that keep one sort of time gradually get out of step with those that keep the other. Shown here is the relationship between the two times. The key feature is that zero for t corresponds to the infinite past for τ.*

astronomical clocks, which determine the Earth day and the year. But "gradually" is the word. From Milne's formula, it follows that the accumulating discrepancy at our epoch is only about one part in ten billion per year. It would take thousands of years for the difference to add up to one second. Therefore, we wouldn't feel any temporal disorientation. On the other hand, if we go back over the past cosmological epochs, the differences mount up, as Fig. 5.2 shows. In fact, as *t* approaches zero, which you might have been fooled into thinking was the beginning of time, τ goes to minus infinity—that is, it stretches into the infinite past. Measured in Earth years, the universe is *infinitely* old! Because Hubble's work on the expansion of the universe also refers to a dynamical rather than an atomic process, it also measures τ time, so on this time scale the big bang happened an eternity ago. Viewed in *t* time, processes such as the rotation of planets were greatly speeded up in the past, but the behavior of light and atomic processes remained unchanged. Conversely, in τ time, the galaxies do not move—i.e., the universe is actually static. Instead, light frequencies gradually decrease with time. This explains the red shift.

Does Milne's theory solve the age-of-the-universe problem? On the face of it, the answer is yes. Milne clearly thought so: "It thus appears that the paradox into which contemporary physics is led in discussing the age of the universe . . . is due to the confusion of the two scales of time," he wrote. However, his theory had very little to say about radioactivity, in particular whether radioactive half-lives refer to t time or τ time. If the latter, then nothing is solved, because the radioactive dating of the Earth is reckoned on the same time scale as the expansion of the universe.

Well, kinematical relativity is only a theory, and you can't find anyone to speak for it today. On the other hand, the idea that there might exist *two* (or more) scales of time in the universe cannot be dismissed so lightly. There is no logical imperative that compels all varieties of clock to agree, and no known law of physics either. Nor was Milne alone in his proposal. No less a physicist than Paul Dirac, one of the founders of quantum mechanics and a Nobel Prize winner, arrived at a similar conclusion to Milne's.

Dirac toyed with the idea of two times briefly in the 1930s, and then put his theory to one side, following which it faded into obscurity. Famous for his diffidence and reserve, Dirac was not one to plug away at an idea unless he had something important to communicate. Although he spent most of his working life at Cambridge University surrounded by some of the world's finest scientists, he rarely collaborated, and his verbal exchanges were monosyllabic. If he thought at all about cosmic time after the publication of his original paper, nobody was likely to know.

When he retired, Dirac moved to Florida, and it was from there that he traveled to Trieste to attend a symposium held in his honor in 1972, on the occasion of his seventieth birthday. I was lucky enough to attend the symposium myself, which was the only time I ever heard Dirac lecture. I well remember the great man rising to give the keynote address. He looked the very model of an English middle-class gentleman, slightly stooped, with gray hair and mustache, and a quietly unassuming manner. The audience lapsed into deferential silence, and I wondered what gems of wisdom he would convey on this very special occasion, after a lifetime quarrying nuggets of truth from the rock face of science. Dirac's low-key style was legendary, and his presentation that day was as throwaway as any I have encountered. After asking for the "lantern" to be switched on, and making the most minimal remarks, he proceeded to show a series of slides. I was staggered to see that after decades of silence on the subject Dirac had chosen to discuss his work on t and τ times! A memorable feature of his slides was Dirac's unconventional practice of drawing time flowing *down* the screen rather than up. The Trieste lecture turned out to be a curtain-raiser on a revitalized research project that occupied Dirac's final years, conducted in collaboration with Vittorio Canuto. What began in the 1930s as a short comment, a sort of half-formulated curiosity, metamorphosed into a full-blooded theory with major ramifications.

Like Milne, Dirac concluded that atomic clocks get out of step with astronomical clocks. As I explained above, viewed in atomic time (the one by which we set our watches, and presumably the one by which our brain activities go), the planets slowly change their orbital speeds. It turns out that a slow variation in astronomical time mimics a slow variation in the strength of the force of gravity between all bodies, so that as time goes on the sun's grip on the Earth and the Earth's grip on the moon gradually weaken. Based on the revamped theory, Dirac and Canuto predicted an alteration in the orbital periods of the planets of a few parts in one hundred billion per year.

By good fortune, it became possible to test for this tiny effect. As I mentioned in Chapter 3, the Mars Viking lander spacecraft provided physicists with an unplanned opportunity to make accurate time and distance measurements across the solar system. The Viking probes were designed to soft-land on the Martian surface and transmit back data on the physical conditions. In particular, they conducted a number of experiments to search for bacterial life. Their relevance to gravitation and Dirac's theory was purely incidental, but felicitous. The problem in testing theories of varying G, as they are known, is that very slow changes in the gravitational force affecting the planets are extremely hard to detect, for two reasons. First, the planets don't merely orbit the sun in pristine Keplerian fashion. They are subject to a large number of small perturbations from the other planets, which all add up to a complicated mess. You need a mammoth computer program to untangle it all, and even then uncertainties remain. Second, to chart the course of a planet through the heavens requires highly accurate position measurements. This was where Viking came in handy. By transmitting from a fixed location on the Martian surface, the spacecraft were able to provide the project scientists with extremely accurate distance measurements. Data were collected over several years and fed into a computer. Meanwhile, Dirac died. Shortly afterwards, Canuto announced that the Viking data ruled out Dirac's theory once and for all!

None of this proves, of course, that there is but one single scale of time for all physical processes; it merely shows that the specific theories of Dirac and Milne are flawed, probably fatally. In the absence of a unified theory of all physical processes based on the notion of a common scale of time, the intriguing question of how many time scales exist remains open. There are many varieties of clock—astronomical, pendulum, atomic, sapphire crystal, superconducting resonators, etc.—involving different physical principles. It is entirely conceivable that some of these clocks can slowly lose their synchrony over cosmological time. Rapid improvements in the accuracy of different sorts of clocks has led to a recent upsurge in intercomparison experiments. For example, a German group has compared a cesium-beam atomic clock with a superconducting cavity resonator over a period of twelve days, and determined that any systematic drift had to be less than about one part in a hundred billion per year. Even stronger limits, of two parts in ten trillion per year, have been placed on the relative rates of magnesium and cesium atomic clocks. Of course, however accurately these experiments are performed, there is always the possibility of still smaller variations.

So where does this leave the problem of the age of the universe? In 1952, the Dutch astronomer Walter Baade shocked his colleagues by announcing that Hubble's results contained a serious error. By then Hubble had been laboring away for two decades measuring the red shifts and distances of ever-fainter galaxies with the aid of the Mount Wilson telescope and his skilled assistant Milton Humason. From the outset, Hubble's method of measuring distances had been based on observations of a particular class of stars known as Cepheid variables. These stars wax and wane in brightness in a distinctive cyclical manner, and measuring the duration of the luminosity cycle allows you to calculate the intrinsic brightness of the star. By comparing its true brightness with its apparent brightness, you can estimate how far away the star is situated. Hubble and Humason sought Cepheid-variable stars in other galaxies to compute their distances. Although the method is a good one, Hubble had been working with an incorrect calibration: the Cepheids were at least twice as far away as he had all along supposed. Suddenly the size of the universe was doubled, and its age stretched by a similar factor. The problem that the Earth seemed to be older than the universe, if far from solved, was at least ameliorated.

Since that sudden rethink, the estimated age of the universe based on Hubble's relationship between the distance of a galaxy and the speed of its recession has been revised upwards several times. Figures of fifteen or even twenty billion years for the age of the universe have been published. For a decade or two, it began to appear that the paradox of a universe apparently younger than some of its component parts had been resolved. Then it all began to go wrong again.

CHAPTER 6
EINSTEIN'S GREATEST TRIUMPH?

I am a detective in search of a criminal—the cosmological constant. I know he exists, but I do not know his appearance.

ARTHUR EDDINGTON (1931)

THE HANDWRITING OF GOD

"Scientists report profound insight on how time began," proclaimed the banner headlines on the front page of *The New York Times*. The date, 24 April 1992, is etched into the memory of every astronomer. All over the world, the press had worked itself into a frenzy about a momentous scientific event. Stephen Hawking called it "the discovery of the century, if not all time." *Time* magazine referred to "Echoes of the Big Bang," and *Newsweek* titillated with "The Handwriting of God."

The hot news was not exactly theological, more cosmological. It concerned a breakthrough in the long-running data analysis from COBE, the satellite designed to search the cosmic background heat radiation for any trace of irregularity. For over two years, COBE had been patiently examining the afterglow of the big bang for hints of any hot spots. As I explained in the last chapter, preliminary observations suggested the background heat radiation is smooth across the sky to at least one part in a hundred thousand. By 1992, enough data had been sifted to discern the presence of a slight but unmistakable pattern in the cosmic heat map. The radiation was unquestionably imprinted with tiny ripples, hot and cold patches superimposed on its otherwise astonishing uniformity. It was exactly what the scientists needed to confirm their ideas about the big bang. "If you are religious," ventured elated project leader George Smoot in an unguarded moment, "it's like looking at God." And the media went mad.

Around the world, scientists breathed a collective sigh of relief. In truth, the big bang theory was in deep trouble, and finding those ripples was crucial. For a while, it had seemed that they might not be there at all. Had that been so, cosmologists would have been obliged to go back to the drawing board.

The significance of the COBE ripples is easily grasped. The cosmic heat radiation has supposedly been traveling more or less undisturbed since about three hundred thousand years after the big bang—the epoch at which the universe had cooled off enough to become more or less transparent. Compared with the present age of many billions of years, that is early. So the radiation is a direct relic of the hot, dense early universe, a sort of snapshot of how the cosmos looked in its primeval phase. Obviously, it was very smooth.

The smoothness of the primeval universe rests uneasily beside its present clumpy structure. Astronomical surveys reveal stars and gas clustered into galaxies, the galaxies congregated into groups, with the groups themselves forming superclusters. During the 1970s and 1980s, astronomers painstakingly mapped the sky in greater and greater detail, building up three-dimensional images of how galaxies are organized on a grand scale. Evidence began to mount of huge voids where luminous matter is rare, enveloped in a ragged patchwork of sheets and filaments formed by thousands of galaxies aggregated together. Vast features stretching for hundreds of millions of light-years across the universe were discerned. This large-scale cosmic texture is reminiscent of the froth on the surface of a glass of beer, or perhaps of a densely woven cobweb.

A major challenge to cosmological theory is to explain how this large-scale structure has arisen. Gravitational forces will naturally tend to pull matter into clumps. If the universe began in a fairly smooth state, with gas spread more or less uniformly through space, then over time there would be a tendency for the gas to become attracted to those regions where the density was slightly higher than the surroundings. As the gas slowly accumulated into blobs, the gravitational pull of these denser regions would be amplified, and the blobs would gather up more and more material at the expense of the rest. Over time, matter would become strongly clustered. When scientists began investigating this clumping process in detail, they soon discovered that it happens only extremely slowly. The problem concerns the expansion of the universe, which works against the tendency for gravity to pull things together. To get the present degree of clustering from completely smooth beginnings would take tens of billions of years—and the universe isn't supposed to be that old.

The same old problem had cropped up once more: there just didn't seem to be enough *time* for an observed feature of the universe to arise by well-understood physical processes. So a way out was suggested. Maybe the universe began with a head start. Perhaps, in the beginning, matter wasn't *completely* smooth after all. Might it have been partially clumped already, so that gravity could finish off the job more rapidly? The snag was, such a hypothesis seemed extremely *ad hoc*. Why should the universe conveniently start out with clumps of the right sizes and densities? To assume that the universe was simply "made that way"— with exactly the right degree of primordial clumpiness—was stretching credulity. A more serious difficulty concerned a conflict with observations. If there *were* clumps in the early universe, then they should show up as distinctive ripples in the heat radiation. But until COBE, this radiation seemed to be completely smooth.

With growing desperation, cosmologists sought a way out. One idea to help alleviate the problem was to invoke the existence of dark matter. Sky surveys map luminous stuff, but matter that doesn't shine will be overlooked. If the universe contains a lot of invisible matter too, the extra material could enhance the gravitational power of the clumps and accelerate the accumulation process. This theory was certainly credible. Astronomers have good evidence for dark matter in the outer halo of the Milky Way, and also within galactic clusters. Some estimates place the amount of dark matter substantially in excess of the visible stuff. Theorists had no trouble drawing up a plausible list of candidates for what the invisible material might be: black holes, dim stars, planets, rocks, neutrinos, unknown subatomic particles coughed out of the big bang. However, it is not enough to throw together a random cocktail of objects and hope for the best: it has to be the right sort of dark matter to do the job properly. We need to explain the observed magnitude of the clumping and its variation over different length scales. For instance, a particular type of dark or invisible matter might cause a lot of clumping on the scale of a few million light-years and very little over billions of light-years; or vice versa. The details need to match.

Astronomers divided their dark matter candidates into "hot" and "cold." Hot dark matter meant light particles, such as neutrinos, which would continue moving very fast as the universe cooled. Cold dark matter referred to heavy things like black holes or dim stars, which would be moving slowly. Computers were used to simulate how the universe would evolve from smooth beginnings in the presence of various sorts of hot and cold dark matter. After many runs, hot dark matter didn't seem to work, so the fashion shifted to cold dark matter. But this didn't fit the bill either. It works well enough on small scales, but less well on large scales. Jim Peebles, the leading Princeton cosmologist, who had a hand in the original discovery of the background heat radiation, was unequivocal: "Cold dark matter is dead," he said.[1] Others began thinking the unthinkable: perhaps there was something fundamentally wrong with the standard big-bang theory?

DID THE BIG BANG EVER HAPPEN?

Astrophysicists of today who hold the view that the "ultimate cosmological problem" has been more or less solved may well be in for a few surprises before this century is out.

JAYANT NARLIKAR

The time-scale difficulty has never been far from the surface in cosmology, and the problems about the slow growth of cosmic structure soon attracted comment from opponents of the big-bang theory. Fixing the date of the big bang hinges, you will recall, on the use of the red shift of distant galaxies to provide a measure of the expansion rate of the universe. The assumption that the red shift provides, on average, a reliable indicator of expansion rate goes right back to Hubble. But could this interpretation of the red shift in terms of the systematic recession of the galaxies be misconceived? After all, other mechanisms are known that produce a red shift, such as the gravitational field of a concentrated mass. And who knows what new physics might apply under exotic and extreme physical conditions?

Some dissenting astronomers have for many years been diligently collecting examples of astronomical misfits, galaxies and quasars that seem to defy the standard interpretation of the red shift. Champion dissenter is Halton Arp of the Max Planck Institute for Astrophysics in Munich, Germany. He has received strong support from British theoretical astronomer Fred Hoyle, he of the old steady-state theory, and Indian and American colleagues Jayant Narlikar and Geoffrey Burbidge. The essence of Hubble's claim that the universe is expanding is the relationship between the distance of astronomical objects and the amount of their red shifts: the distant objects have the higher red shifts, in exact proportion. The validity of the Hubble "law" hinges on the existence of a good, dependable method for determining distances. For nearby galaxies, astronomers can pick out cepheid variable stars to provide an accurate yardstick, but more distant galaxies are too faint to allow this. A rough guide is provided by the apparent brightness of the object. Obviously, the farther away a luminous body is located, the dimmer it will seem from Earth. But for this method to work properly, you need to know the intrinsic brightness of the body to start with. If an object is intrinsically dim, there will be a tendency to overestimate its distance.

Astronomers developed statistical techniques to avoid this sort of bias. In the case of normal galaxies, which are fairly well-understood objects, the results look fairly sound. Then, in the 1960s, new classes of objects were discovered, such as the intensely bright quasars, or QSOs, and galaxies with their nuclei strongly disrupted by highly energetic processes. Since these objects were found to have very high red shifts, most astronomers concluded they were situated at vast distances, "on the edge" of the visible universe. On the other hand, as nobody knew their intrinsic brightness, there was no easy way to determine the distances, so the tidy relationship between distance and red shift was missing.

By the early 1970s, Arp, Hoyle and company had begun openly questioning whether the red shifts of these unusual objects are in fact produced by recession. Their challenge was based on the discovery that many high-red-shift QSOs are situated in the sky in close proximity to galaxies having much lower red shifts. If two objects with very different red shifts are located side by side in space, then Hubble's law is discredited, and the whole basis of modern cosmology, including the expansion of the universe and the date of the big bang, disintegrates. With stakes as high as this, it is small wonder that cosmologists reacted coolly to Arp's claims. The true explanation for the discrepant red shifts lay, they countered, in chance alignments. Given a random distribution of objects in three-dimensional space, it is to be expected that, here and there, some faraway object will be positioned in the sky close to a much nearer object, in the same way that a foreground tree can align with a distant mountain if viewed from a particular perspective. With so many galaxies, it is bound to happen sometimes. So the issue descended into a squabble over statistics. Just how probable is it that a random selection of galaxies and QSOs will show x chance alignments? How likely is it that astronomers subconsciously select aligned objects from a scattered field? Those on both sides of the debate still remain convinced of their positions twenty years on.

In 1971, Arp strengthened his argument by finding a pair of objects—one a QSO called Markarian 205, the other a spiral galaxy, NGC 4319—that seemed to be connected by a faint bridge of light. Spectroscopic analysis, however, showed the quasar to have a much bigger red shift. If the usual interpretation is made of identifying red shift with recession, then the galaxy is receding at 1,700 kilometers per second, the QSO at a snappy 20,250 kilometers per second. Arp argued that the bridge of luminous matter connecting the two objects proved they were located side by side in space, and suggested that the QSO has somehow been ejected from the galaxy, leaving a kind of trail.

Other examples were found. One of these, known as Stephan's Quintet, is a very tight cluster of galaxies in which the objects look as though they are disturbing each other gravitationally, something that can happen only if they are cheek by jowl in space. Yet the red shifts suggest recessional velocities ranging from 800 to 6,700 kilometers per second. In another case, a luminous bridge joins the galaxy NGC 7603—apparent recession velocity 8,800 kilometers per second—with a smaller high-red-shift neighbor—apparent recession velocity 16,900 kilometers per second. Then there are three QSOs aligned close to the center of the spiral galaxy NGC 1073, and another three close to NGC 3842.

Arp and his colleagues insist these associations are physical and not merely fluky geometrical alignments. They claim the QSOs are being flung from relatively nearby galaxies, and that their red shifts cannot be fitted into the Hubble scheme and are not due to the expansion of the universe. To support the ejection hypothesis, they point to some cases in which QSOs are strung out in straight lines, or are aligned parallel to jets from associated galaxies. In a recent series of papers, they propose an abandonment of the big-bang theory altogether, and a return to a variant of the old steady-state model, in which there is no origin of time.

Defensive astronomers have tried countering these ideas by appealing to a phenomenon known as gravitational lensing. One of the key predictions of Einstein's general theory of relativity is the bending of light by a massive body. As mentioned in Chapter 4, Einstein predicted that the sun should slightly bend starbeams, an effect confirmed by Eddington in 1919. Shortly afterwards, Sir Oliver Lodge pointed out that, if a light source lay exactly behind a massive object, then light from the distant source would be bent around all sides of the interposing object and focused along a line (see Fig. 6.1). An observer located on the line would see a bright ring of light surrounding the occulting object, known as an

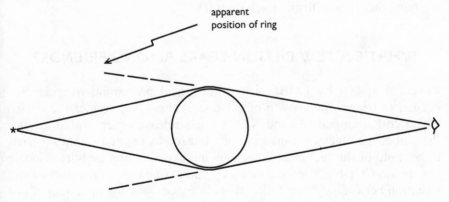

6.1 Einstein ring. The light from a distant star can be focused by the gravity of an intervening massive object (star or galaxy), and appears to an observer as a ring of light.

"Einstein ring." In the 1930s, Einstein investigated this idea himself but dismissed it as of theoretical interest only, believing there was no hope of ever detecting it. He was wrong. Examples are now known of distant galaxies and QSOs being lensed by nearby galaxies, producing multiple images, and in some cases creating all or part of an Einstein ring. The focusing effect also makes the distant light source seem much brighter.

Gravitational lensing can be produced by any object, from galaxies down to dwarf stars, planets and even asteroids. In 1993, a team of American and Australian astronomers at Mount Stromlo Observatory in New South Wales reported the observation of an unusual lensing event. It was caused by an unseen dwarf star in the halo of our galaxy interposing itself between Earth and a normal star in the Large Magellanic Cloud. The astronomers saw the star wax in brightness for several days. They conjectured that there are many such unseen stars in our galaxy and others, contributing to the dark matter of the universe that I mentioned above. From time to time, an unseen star will also lens a very dim and distant QSO. When this happens, the QSO will appear much brighter, giving the impression it is quite close to us. This could account for some of Arp's misfits. But it is not clear that such effects will explain all the associations between galaxies and QSOs, still less the existence of "bridges" between objects of differing red shift.

While the debate over discrepant red shifts rumbled on, other odd bits of evidence kept cropping up that challenged the orthodox big-bang theory, such as the discovery of more objects that seemed to be older than the universe, and some curious observations that suggested a large-scale periodicity in the distribution of galaxies. The accumulating difficulties prompted American physicist Eric Lerner to write a book provocatively titled *The Big Bang Never Happened,* published in late 1991. A few months later, the COBE ripples were discovered, and suddenly the big-bang theory was firmly back on track.

WHAT'S A FEW BILLION YEARS AMONG FRIENDS?

Revealed at last by COBE were the crucial primordial irregularities needed to trigger the growth of galactic clumps. No wonder the champagne corks popped. As the NASA press release reverberated around the globe, the electronic-mail circuits buzzed as excited scientists tried to get hold of the technical data. Key indicators would lie buried in the fine detail. COBE garnered its data by comparing the temperature of the radiation between different directions in space, building up a map of hot and cold patches in the sky. The cause of the temperature variations is the gravitational red-shift effect created by the gravity of the clumps of gas. The cold patches in the sky are therefore effectively timewarps created by colossal aggregations of primordial matter. Very soon cosmologists were referring to the ripples as "wrinkles in time."

It was clear from the start that the magnitude of the temperature variations (about thirty-millionths of a degree) was on average the same over the entire angular range surveyed—about nine degrees upwards. This carried an important implication. No scale of size was apparent: big-sized ripples and little-sized ripples were equally pronounced. The scale-independence of the ripples delighted many theorists, because it exactly fits what they expect from their favorite variant of the big-bang theory. Known as the "inflationary-universe scenario," it postulates that just after the big bang the normal gravitational braking effect (recall Fig. 5.1) was briefly interrupted, and the universe abruptly jumped in size (inflated) by an enormous factor. One consequence of inflation would be to iron out any initial irregularities, on all length scales, leaving the universe pristine smooth. The ripples found by COBE would presumably have been added after inflation, perhaps as a result of quantum fluctuations, and should have no preferred length scale—exactly as COBE found.

However, the absence of a length scale carried with it another implication, this time concerning the age of the universe. Here the message wasn't so good. In Chapter 5, I explained how Einstein's field equations predict many possible cosmological models. The job of the cosmologist is to try and use the observational facts to select a particular best-fit model. The various models on offer differ in the precise manner in which the universe expands from its big-bang origin. The general trend is shown in Fig. 5.1, which shows how the size of a typical region of the universe grows with time. Notice how the curve starts out vertically at time zero, corresponding to the explosive origin, and steadily bends over as gravitational braking slows the expansion. The expansion rate today, deduced from red-shift data, is given by the slope of the curve at the present epoch, marked "now." The age of the universe is the distance along the time axis from the start of the curve to "now."

To estimate the age of the universe, you need to know two things: the present rate of expansion, and the size of the braking effect. To see why *both* quantities are needed, look at Fig. 6.2. Curve A shows a weakly braking universe, curve B a strongly braking version. Notice that the extra deceleration in B shows up in the way the graph curves over more sharply. For a given value of the slope (i.e., the expansion rate), it is clear that curve B predicts a universe much younger than curve A. The strength of the braking is determined by the amount of matter in the universe. The more there is, the stronger its gravitational pull and the more the expansion rate decelerates. As I have mentioned, the amount of matter in the universe is a bit of an unknown. There is clearly a lot of dark matter, but astronomers can't agree on exactly how much.

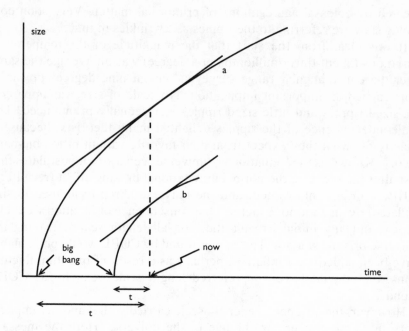

6.2 Braking the universe. Curves a and b compare the behavior of universes with a low and high matter density respectively. Model b decelerates more strongly, due to the higher gravity. For a given slope of the curve now (determined by the present expansion rate), the time t that has elapsed since the big bang is greater in a than b.

Hidden in these graphs is the story of a gargantuan tussle between the force of gravity, which tries to pull the retreating galaxies back, and the impetus of the big bang. Gravity weakens its grip with distance, so, the more the universe expands, the less braking there is. On the other hand, the rate of expansion is slowing all the time. If there is enough invisible matter in the universe (about a hundred times the amount of visible matter), the combined gravitational attraction will eventually halt the expansion completely and turn it into a collapse. On the other hand, if there is rather less matter, then the battle will be won by the expansion: at some stage, the galaxies will essentially free themselves from gravitational restraint and expand more or less freely thereafter. Either way, there will be a "moment of truth" at which the outcome is decided. In the case of recollapse, this will be the epoch at which the universe reaches its maximum size. If the expansion continues forever, it will be when the braking effect becomes negligible.

The existence of a particular "moment of truth" introduces a characteristic time into the description of the universe—the time of decision.

Associated with this time will be a characteristic scale of length: the distance that light has traveled since the big bang up to that special epoch. But if the universe has a special length scale built into it, that length ought to show up in the COBE ripples at the corresponding angular size. Yet no such length scale is observed: the degree of irregularity is the same on all scales. Why?

One answer immediately suggests itself. The two scenarios just summarized—free expansion and recollapse—merge in a limiting case for which the gravitational pull declines in exact proportion to the diminishing impetus of the expansion. In other words, the above-mentioned tussle is evenly matched. If the universe is like this, the battle will never end, meaning that the universe will expand forever, but at an ever-diminishing rate. This halfway house was, incidentally, the model that Einstein eventually plumped for after he learned of the cosmological expansion. Today it is called the "Einstein–de Sitter universe." Because the battle is endless, the "moment of truth" is postponed indefinitely, so there is no characteristic time scale or length scale in the theory. The Einstein–de Sitter model predicts that the COBE fluctuations should be scale-independent, exactly as observed. Fortunately, the Einstein–de Sitter model is also required by the simplest version of the inflationary theory.

But now we hit a snag. One problem about the Einstein–de Sitter model is that it has a lot more braking than most astronomers would like. Consequently, the inferred age of the universe becomes uncomfortably short once more. If we take a plausible value for the expansion rate of the universe, the time since the big bang works out at only about ten billion years. But as I already explained in Chapter 5, stars are known with ages of at least fourteen billion years. Recent observations have actually increased this figure: it is likely that stars near the centers of large galaxies are one or two billion years older than those in globular clusters. Sometimes ages as high as seventeen billion years are quoted, and a recent claim from California suggests that a star with an age of nineteen billion years has now been discovered. Clearly something is wrong.

The age problem derives from the fact that the rate of expansion of the universe is "too" high: the faster the universe expands, the more recently it must have been in its compressed big-bang state. The rate is expressed in terms of a speed divided by a distance. Hubble himself gave the value of 540 kilometers per second per megaparsec. (A parsec is an astronomical unit of distance, equal to 3.26 light-years.) This figure means that a galaxy 10 megaparsecs away recedes at 5,400 kilometers per second, whereas one that is 100 megaparsecs away recedes at 54,000 kilometers per second, and so on. Except that Hubble got his sums wrong, as we have seen. A skillful and devoted student of Hubble, Allan

Sandage, has spent his life measuring the expansion rate. Sandage is considered by many to be *the* senior American astronomer and Hubble's natural heir. For many years, he has quoted the value of 50 kilometers per second per megaparsec. Unfortunately, another group of astronomers, led by French-born Gérard de Vaucouleurs of the University of Texas at Austin, hotly disputes this figure, and quotes the much higher value of 100. The difference here is crucial. If 50 is correct, the age of an Einstein–de Sitter universe is about 13 billion years. Perhaps the astronomers are a bit out in their estimates of the ages of the oldest stars, and they just fit into the time available? But if 100 is correct, the universe would be a mere 6.5 billion years old and the inconsistency is glaring.

Curiously, until recently few astronomers have felt inclined to take the average of the Sandage and de Vaucouleurs values, preferring to join one camp or the other. But a number of careful analyses of the data have now come out with values around 70 or 80. This still does not jibe comfortably with the ages of the stars if the Einstein–de Sitter model is correct. (For 80, the age works out at a little over 8 billion years.) Once more, the absurd conclusion seems forced on us that the age of the universe is less than the ages of some of its parts.

What does this mean? Some cosmologists claim it casts doubt on the entire big-bang theory. Its failure in such a crucial test as this is, they say, decisive, and leaves the door ajar for a fundamental reappraisal of physical cosmology. Are those discrepant red shifts real after all? Perhaps cosmic time *is* different from Earth time, as Milne long ago suggested? Maybe the big bang didn't happen after all, and the universe is infinitely old?

These rebels are in a small minority, however. Most scientists prefer to wait and see if the expansion-rate or stellar-lifetime data get revised. Others reject the inflationary scenario, and think it too early to conclude much from the COBE data. They want to wait until this information is supplemented by ground-based observations of ripples on smaller angular scales before making a judgment. But there is an unmistakable air of discomfort surrounding the age problem. Astronomers clearly prefer not to think about the fudge and fit needed to squeeze fifteen billion years into ten. Commenting recently on the time-scale discrepancy, Arp, Hoyle and colleagues observed: "For some reason it is not being discussed, but in terms of numerical factors the problem for the Big Bang is back again."[2]

There is, however, a tidy way to retain the inflationary-big-bang theory *and* square all the numbers, comfortably accommodating the COBE data and the more embarrassing values of Hubble's constant. This felicitous mix can be achieved by turning Einstein's greatest mistake into his greatest triumph.

A REPULSIVE PROBLEM

Away with the cosmological term . . .

ALBERT EINSTEIN

In Chapter 5, I explained how Einstein, having produced his beautiful gravitational-field equations in 1915, then "sullied" them by adding an extra term—the cosmological or Λ term. He came to regret this step bitterly. First, it snatched away the chance for him to predict the expansion of the universe. Second, the extra term reeked of fudgery. Indeed, Λ has become known as "Einstein's fudge factor," unworthy of a theory so breathtakingly elegant and awesomely powerful as general relativity, and unbecoming of a man of such purist tastes.

Taking their cue from the great man, scientists in general have tended to regard the Λ term as being as repulsive as the force it describes. Partly this is a reaction to Einstein's dramatic U-turn, partly it is because of Ockham's razor. Why add an extra term to a set of equations that is already demanding enough? It only serves to multiply the choice of cosmological models on offer and obscures the interpretation of the astronomical observations.

There is another reason why scientists would rather set Λ equal to zero. Cosmological observations limit the size of Λ to at most a very small value. As I described in Chapter 5, a Λ force is exceedingly feeble by any standards, orders of magnitude weaker than anything else known. Many physicists dream that the various forces of nature—gravitation, electromagnetism and the nuclear forces—will one day be combined in a unified field theory of the sort that Einstein heroically sought to construct in his latter years. It is hard to see how such a theory would predict one force so much weaker than all the rest.

Stephen Hawking has presented an elegant argument along these lines.[3] To quantify just how weak the Λ force is, we must compare it with something. A convenient way to do this is in terms of the range at which the force makes itself felt. As explained, the Λ force is certainly negligible at distances of less than a few billion light years. The weaker the force, the greater the range at which it manifests itself. If the force is zero, the range is infinite. One can also discuss the range of the more familiar electromagnetic force, but in this case things are "back to front." As I already pointed out, the Λ force is unusual because it grows *stronger* with distance rather than weaker. By contrast, the electromagnetic force declines with distance, so a measure of its range is the distance *beyond* which it becomes negligible.

Observations of the magnetic fields of galaxies suggest that electromagnetic effects extend over at least a million light-years, but little is known about distances greater than this. It is *possible* that the electromagnetic force abruptly vanishes at, say, a billion light-years from its source, but almost no physicists believe this. They argue that, as the range is already known to be so great, it really should be *infinitely* great, for it is hard to imagine how a fundamental distance as large as a billion light-years could enter into the basic laws of electromagnetism. Rather than simply concede that the range of the electromagnetic force is some unknown quantity greater than a million light-years, physicists appeal to a mathematical-symmetry principle (called "gauge symmetry") that *fixes* the range to be truly infinite. This beautiful symmetry, implicit already in the electromagnetic equations of Maxwell, also serves to make electromagnetism simple and elegant. Now, compare the situation of electromagnetic theory with the cosmological repulsion. The range of the Λ force is known to be much greater than a million light-years. If the same reasoning as above is adopted, we ought to argue that this range should be infinite, and seek a deep mathematical-symmetry principle—akin to electromagnetic gauge symmetry—that automatically fixes Λ to be *strictly* zero. Unfortunately, nobody has yet discovered what this symmetry is.

Another argument for a strictly zero value of Λ comes from a study of theories which attempt to unify the nuclear and electromagnetic forces. Even though these other forces have no direct connection with gravitation, nearly all the unified theories contain physical processes that mimic a Λ force. In other words, a cosmic force appears as an inescapable by-product of the other forces of nature. The trouble is, the magnitude of this by-product is truly stupendous—typically 120 powers of ten larger than the value cosmologists have looked for. If such a force existed with this enormous strength, it would blow the universe apart in less than a microsecond!

The existence of a huge cosmic force in these unified theories is a severe embarrassment. One suggestion for how it may be circumvented is to suppose that many different physical processes act to create Λ-type forces, but some of these produce negative values for Λ. It is then possible to imagine a cancellation between negative and positive contributions, yielding zero for the net value. This type of exact nullification is known to happen in physics. For example, there is a deep underlying symmetry in nature for electric charge: for every positive charge in the universe, there is a negative charge to balance it. It is not too hard to believe that a symmetry of this sort applying to Λ lies buried in the basic laws of nature. Much harder to swallow, however, would be the hypothesis that this cancellation of positive and negative values is almost but not quite exact, so that a tiny positive quantity is left. Arithmetically, it implies that the positive and negative terms neutralize each other to the staggering fidelity of 119 powers of ten, but fail to balance perfectly by a mismatch as small as one part in 10^{120}. Can we really believe that nature would do this?

In attempts to make Λ go away, theoretical physicists have wrestled with a variety of ideas. One of these is to treat the Λ term as if it described a sort of field in its own right, with its own dynamics. Applying quantum mechanics to that field has led to the conclusion that the most probable value is indeed very close to zero.

Suggestive though these various calculations and arguments might be, they have not won universal acclaim, and physicists place "the cosmological-constant problem" high in the list of outstanding scientific mysteries. Why is it a problem? Because most physicists would rather not have a cosmological constant, but lack a convincing scientific argument to set Λ equal to zero. In the absence of an argument that Λ *must* be zero, we certainly cannot rule out a cosmological term. The American physicist and cosmologist Steven Weinberg points out that nature has a tendency to bring into being all those things that are not specifically forbidden by a symmetry principle or other form of law. "There is no reason *not* to include a cosmological constant in the Einstein field equations," he maintains.[4]

THE LOITERING UNIVERSE

Not all cosmologists have despised the cosmological constant. Eddington, for one, positively welcomed it. Together with Georges Lemaître, he proposed a big-bang model of the universe containing a Λ force. The cosmological term makes virtually no difference in the compressed early phase of the universe, because the Λ force is very weak at short distances. However, as the universe expands, so the repulsion gains in strength, and this has the effect of counteracting the normal attraction of gravity. Just as normal gravity acts as a brake on the expansion, so the Λ term acts rather like an accelerator, forcing the universe to expand faster. At first the braking effect dominates, so the expansion rate decelerates as usual, but as the universe grows larger, so the competition between these opposing forces becomes more evenly matched. Eventually a stage is reached at which the forces neutralize each other, and the universe hesitates, unsure of whether to speed up or slow down. The result is that it starts to "coast"—i.e., expand at a nearly uniform rate. This loitering phase cannot go on indefinitely, though, as the continued expansion will eventually give the repulsion the upper hand. Slowly but surely, the universe starts to accelerate in its expansion, with the Λ force strengthening all the time. The situation that pertains near the big bang is therefore reversed, with the attractive force of gravity fading away to

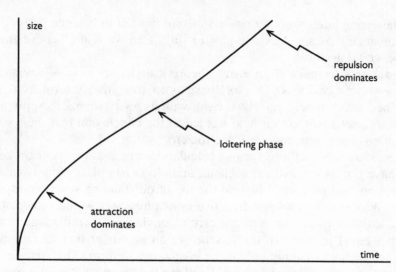

6.3 *Loitering universe. This model, proposed by Eddington and Lemaître, includes the effect of a cosmological-repulsion term. The universe decelerates slowly, and after a phase of almost free expansion ("loitering"), embarks on accelerated expansion.*

leave a universe dominated by repulsion. As a consequence of this runaway victory, the universe goes on expanding forever, growing larger and larger at an accelerating pace.

The general behavior of the Eddington-Lemaître model is depicted in Fig. 6.3, which should be contrasted with the orthodox Einstein–de Sitter model shown in Fig. 5.1. Tracing the curve from the left, we see how the universe grows in size throughout, but the graph starts by curving to the right and finishes by curving to the left. In between, there is an approximately straight section—the loitering phase. The duration of this phase depends on the value chosen for Λ, but it can be contrived to be as long as we please by making a careful choice.

The distinctive wiggle in the curve of Fig. 6.3 holds the key to the model's appeal, for here lies a ready-made solution to the notorious time-scale problem. This is immediately obvious in Fig. 6.4, a highly exaggerated case in which the cosmological constant has been fine-tuned to match the gravitational attraction to high precision, producing a very long loitering phase. In effect, the loitering approximates Einstein's original static-universe model, sandwiched between periods of decelerating and accelerating expansion. Clear from a glance is that a line with a given slope can be matched to Fig. 6.4 at *two* possible positions, in contrast to the case for the Einstein–de Sitter model. The time that has elapsed since the big bang is obviously much greater if the slope is fitted in the right-hand position than the left. If we assume a sufficiently long loitering period, the age of the universe can be extended indefinitely.

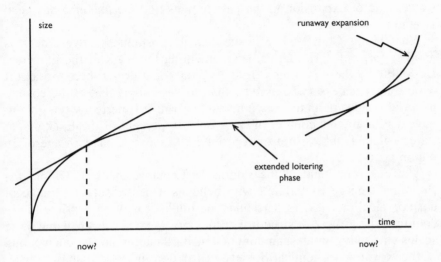

6.4 *When the gravitational attraction and cosmic repulsion are more evenly matched, the universe lingers longer in its coasting phase. An observer measuring a given rate of expansion (slope of the curve) might be located at one of two epochs, leading to greatly different estimates for the age of the universe.*

However, the curve needs to be consistent with observations. During the loitering phase, the universe expands little. Translated into red shifts, this implies a bunching up of values in the middle range. In the 1960s, astronomical surveys of QSOs did indeed hint at such a bunching, but better data removed the effect. More critically, observations of rare instances where distant QSOs are gravitationally lensed by nearer galaxies can be used to place stringent limits on how recently the universe can have loitered. The results of these studies indicate that any loitering must have taken place at a very early epoch, when the Λ force was still very weak. But because loitering needs matched repulsion and attraction, the attractive force of ordinary gravity must also have been very weak—i.e., there would have to be very little matter in the universe —much less in fact than observations suggest. Therefore, it is unlikely that our universe has spent a substantial time loitering.

Nevertheless, the existence of a cosmological constant will always serve to increase the age of the universe somewhat, over and above the age predicted by a model having zero value for Λ. This is so with or without a distinctive loitering phase in the past, because of the accelerative tendency. The reason is easily understood. To reach its present size and expansion rate, the universe had to expand fast in the past, to overcome the braking effect. If the braking was less, the universe could have got to its present state with a less rapid past expansion. But a slower rate of expansion in the past implies that the universe has been around for longer.

What about COBE? The cosmological constant effectively acts as another form of dark matter, supplementing the mass of the universe. Undoubtedly there is some "ordinary" dark matter out there too, but it is no longer necessary to assume that at least 90 percent of the cosmic material resides in an unknown invisible form. It is perfectly possible to have a total matter content of, say, only 10 percent of the Einstein–de Sitter value, Hubble's constant around 80, and still have an age of 16 billion years.

A strong proponent of the Eddington-Lemaître model is Oxford astronomer George Efstathiou, who believes it neatly solves a range of cosmological puzzles. In particular, he thinks it may be possible to account for the growth of structure in the universe on both small and large scales with only a modest amount of cold dark matter alone. At a meeting of the Royal Astronomical Society in London in 1993, Efstathiou presented the results of detailed observations, both ground-based and from satellite, and matched them with computer models of the growth of clumping with various sorts of dark matter. He demonstrated how models with a Λ term can be made to fit all the data snugly.

Although the cosmological term remains the neatest way to solve the cosmic-age problem, it is too soon to give a definite yes or no to the concept. However, it may be that an answer is not long in coming. In 1990, a group of Japanese astronomers and Edwin Turner of Princeton University independently hit upon a new way to measure Λ using gravitational lensing of QSOs. Because a universe with a cosmological term is older, light from distant QSOs will have been traveling longer, and thus will stand a greater chance of passing near an intervening galaxy and being lensed. So a count of the number of lensing events across the sky can be used to place a limit on the size of Λ. My bet, for what it is worth, is that the observations will eventually demonstrate the existence of a cosmological term. It will surely represent the supreme irony: from a study of Einstein rings, which Einstein himself never believed were observable, astronomers would have shown that Einstein's greatest mistake was in fact his greatest triumph.

In October 1994, sensational new results from the Hubble Space Telescope were announced, yielding a Hubble constant of about 80. Some commentators inferred an age of the universe of only eight billion years. The discrepancy with stellar ages is now glaring, and the cosmic age paradox has forced its way back onto the scientific agenda. While some cosmologists began questioning the standard big bang scenario, Hubble team member Barry Madore suggested that the cosmological term could stage a comeback. He told the *Boston Globe,* "Einstein had the answer in his hands when he first formulated general relativity."

CHAPTER 7
QUANTUM TIME

Einstein said that if quantum mechanics is right, then the world is crazy. Well, Einstein was right. The world is crazy.

DANIEL GREENBERGER

TIME TO TUNNEL

The 1,024-node CM5 computer at The University of Adelaide, allegedly one of the fastest in the world, can perform 59.67 gigaflops flat out. That's much faster than the speed of thought (for a human). The CM5 may not be able to outcompose Mozart, fall in love or even be *aware* it is a whiz at computation, but it can certainly do some snappy arithmetic.

Stripped down to its bare essentials, a computer is a vast network of switches and wires configured to perform many simple tasks quickly. Tiny electric pulses course through the unseen circuits, frenetically trading information. Complex patterns of electrical activity weave themselves through the hardware, and a myriad micro switches flip and flop in silent obedience, marshaled by the iron rules of logic.

In their insatiable search for greater and greater computing power, scientists have made ever-faster circuits and switches. Increasingly they are turning from electronics to photonics—the use of light in place of electricity—for extra swiftness. But sooner or later they will come up against nature's own fundamental limitations on speed. Einstein's time prevents any information from traversing the circuitry faster than light. For a computer one meter in size, this places a limit of three nanoseconds on the speed of information transfer across the machine. To circumvent that, computer scientists have made their components smaller and smaller. But now we run into another basic limit: quantum physics. Individual electrons and photons in a computing machine are subject to Heisenberg's uncertainty principle, which introduces an irreducible fuzziness into the very notion of speed, rate and time.

To appreciate the thorny character of the problems involved, consider one of the more bizarre quantum processes widely used in practical electronic devices. It is called the "tunnel effect." Imagine throwing a stone softly at a window. You expect the stone to bounce back again. Suppose that, instead of the stone rebounding from the window, it passed right through and appeared on the far side, leaving the window intact! Anyone seeing a stone penetrating a window without breaking it would conclude a miracle had happened, but this particular miracle occurs all the time in the subatomic domain, where quantum rules defy common sense.

At the atomic level, the role of the stone is played by that of a quantum particle—say, an electron or a photon—and the window might be some sort of thin barrier, perhaps a wafer of material or just an invisible force field. A particle approaching such a barrier without enough energy to break through will nevertheless often be found on the other side, having apparently "tunneled" through the barrier. (See Fig. 7.1.)

Heisenberg's uncertainty principle offers a clue to the tunneling trick. As recounted in Chapter 3, the energy of a quantum particle can't be measured with total precision at a specified instant. Uncertainty in energy can be traded with uncertainty in time, but you will never eliminate both indeterminacies simultaneously: nature will not permit us to know everything about a quantum particle at once. A crude but helpful way to think about this smearing of energy and time is to imagine that the particle is magically able to change its energy (within a strictly circumscribed range) for a short duration. In effect, the energy of the particle can jump about spontaneously within the limits set by the uncertainty principle. Sometimes it is said that the particle can "borrow" energy for

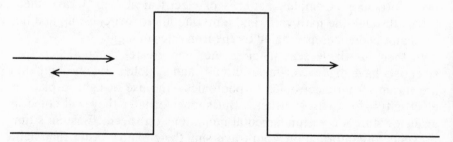

7.1 *Tunneling through a barrier. A stream of quantum particles from the left meets a barrier. Some particles bounce back, but others materialize, magically, on the far side of the barrier, and continue on their way.*

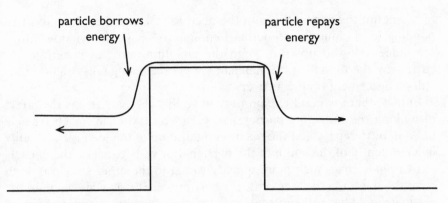

particle borrows
energy

particle repays
energy

7.2 Borrowing energy. According to a loose interpretation of the Heisenberg uncertainty principle, a quantum particle may "borrow" energy to help it get over the barrier. The energy must be repaid promptly on the remote side. This is a helpful way to think about the tunnel effect.

this fixed term. The main point to note is that, the shorter the term of the loan, the greater is its permitted size; an electron, for example, can borrow a lot of energy for a very short time, or a little energy for somewhat longer.

To explain the tunnel effect using energy-time uncertainty, we suppose that the particle is permitted to "borrow" the energy for the purpose of overcoming the barrier. Armed with this bonus, the particle can then advance without hindrance, in the same way that an enfeebled hiker ski-lifted to the top of a mountain can traverse the summit (see Fig. 7.2). The snag is, the quantum particle is living high on strictly borrowed energy. If it can't get to the other side of the barrier and down again before the loan is summarily recalled, it will be obliged to turn back. Such particles simply recoil off the barrier, having penetrated a short distance into it. Although the *maximum* duration of the loan is laid down by law, the specific arrangements for each individual particle are decided by a sort of natural lottery, and may fall short of what is needed to beat the barrier. The process is therefore fundamentally statistical in character. Only a certain fraction of the particles gets through, and it is not possible to know in advance which ones will make it and which ones will have to turn back. Part of the deal, though, is that, the wider the barrier is, the less success the particles will have in penetrating it—i.e., the larger the fraction of particles that gets bounced back.

When the tunnel effect was first discovered several decades ago, an obvious question was: how *long* do the particles take to tunnel through the barrier? You might imagine that the presence of a barrier would slow the particles up. This would be bad news for high-speed computing. Certainly, if the particles are fired with enough energy in the first place to surmount the barrier without the need for a Heisenberg loan (corresponding to pitching the stone hard enough to smash the window), they are indeed slowed up. But when the quantum particles penetrate the barrier by the tunneling mechanism, simple reasoning fails, and there is still no agreement on the answer.

In fact, there is good reason to believe that the barrier has the effect of making the tunneling particles travel *faster*. After all, each particle has a debt to repay, and the maximum duration of the loan is fixed quite independently of the width of the barrier. For wide barriers, the particle has to move that much more quickly to get to the other side in time to clear the debt. On the face of it, if a barrier were made wide enough, the particle would have to move faster than light to penetrate it in the allotted time. Now, that is an intriguing possibility. Unfortunately, the fraction of particles that penetrate barriers of this sort of width is extremely small. Just when the situation gets interesting, the flow of tunneled particles dwindles to a trickle.

Nevertheless, it seems it must be possible to determine how long a particle that *does* tunnel through, even against huge odds, takes to do so. Textbooks on quantum mechanics give a variety of answers. According to some authors, the process is instantaneous: the particle simply disappears from one side of the barrier and instantly reappears on the far side. Others say the time is simply not defined—we can never know the answer. But a computer scientist can certainly determine how fast the machine will compute. That is an observable—indeed, a practical—matter. A lot of money could hinge on beating any apparent quantum limit.

WATCHED KETTLES

Further hints come from other quantum processes. When an atom is excited, an atomic electron jumps to a higher energy level. It remains there for a certain duration before dropping back down to its ground state—a process known as "decay." The excess energy from the decay process is divested, usually in the form of a photon, which flies away. By detecting and measuring the energy of the photon, we can figure out the difference in energy of the atom's levels.

How long the electron spends in an excited state varies from case to case, but a definite prediction can be computed using quantum mechanics. However, an essential feature of quantum physics is its indeterminacy: the behavior of *individual* systems cannot be predicted. The theory can yield the *average* lifetime of the excited state, but it cannot tell you in a particular case precisely when that specific atom is going to decay. This inherent fuzziness prevents us from giving a meaningful answer to the apparently straightforward question: how long does the electron take to jump from one level to another? Try as you may, you will never spot the electron in the act of jumping, or hovering halfway between levels. There is a certain well-defined probability that, after a certain time, the electron will be back in its ground state, having decayed at some unspecified time beforehand. More than that we simply cannot say.

That's nonsense, interrupts our sharp-witted skeptic. Why not observe the atom continuously, armed with a stopwatch, and catch it in the act of decaying?

Good idea! Remarkably, the technology exists to do more or less that. Single atoms may now be trapped, slowed and stored for long periods using electromagnetic fields, then probed with lasers. Unfortunately, even these clever tricks can't outmaneuver the Heisenberg smoke-screen. You can never detect the atom in the process of decay. What actually happens, if you look at the atom closely and continuously, is that the very act of observation itself interferes with the decay processes, and effectively freezes the atom in its tracks (i.e., in the excited state). This phenomenon has been dubbed "the watched-kettle effect," because it is reminiscent of the proverb that a watched kettle never boils. The watched-kettle effect cannot be evaded: to monitor any quantum system, you must interact with it somehow, and this interaction will unavoidably disturb the process under investigation. But take your eyes off the ball for a moment, and the atom infuriatingly decays when you weren't looking!

But surely the decay really did take place at a specific moment when you weren't looking? I can accept that the behavior of excited atoms is somewhat uncertain. But there must be a definite moment, after a duration spent dithering, when the electron makes up its mind and starts to jump—an instant of time when the electron leaves the excited energy level and begins its journey to the ground state? And this journey must itself take a definite length of time. It may be frustrating that a human being can't see it happen, but that is hardly relevant.

It is not only us who are stymied. The Heisenberg uncertainty principle prevents *any* system of apparatus, or indeed any observer, from determining the moment and duration of decay. It is a fundamental limitation on knowledge inherent in the laws of nature, not merely some sort of human failing. However complicated you make your apparatus, you will never be able to snatch a peek at the atom decaying. Einstein spent a long time trying to dream up tricks to find a way around this and eventually gave up the task as hopeless.

Are you saying that the atom doesn't decay at a particular time, or that it does but we can never determine when?

In quantum physics, you have to be clear about what is being measured or observed and stick to that. You can't say much about things that are *not* being observed. In the case of time, we have double trouble, because we never actually measure time as such (in any objective sense). We don't gauge a duration by comparing it mystically with some separate entity—"time"—that looms over all activity with inbuilt "notches" to tally against. If you want to measure time, you have to specify some sort of clock that will do the measuring, and then make an observation of the clock. But a clock is a physical object that changes, and we measure time by observing the *spatial position* of some clock variable, such as the hand. When we say, "Earth takes twenty-four hours to rotate once," we *really* mean, "If the hour hand points to the twelve when the Earth has a certain orientation relative to the sun, then, when the Earth next has that orientation, the hour hand will again point to the twelve"(having been around twice, of course).

Very well, then. Design a clock that will stop the moment the atom decays!

Unfortunately, nature craftily outsmarts us again. There is no such thing as a perfect quantum clock. All real clocks, made of real matter, suffer from the same uncertainty, the same quantum fuzziness, as everything else. When you couple a fuzzy quantum clock to a fuzzy excited atom, you get at least the same degree of fuzziness as always—and you still can't say when the specific atom decayed or how long the process took. Nature seems to have an inbuilt censorship process that always prevents us from knowing precisely when things happen, however devious we may be in our strategy.

ERASING THE PAST

Even God cannot change the past.

AGATHON

Experiments in which attempts to determine the exact moment when the quantum system "made up its mind" yield surprising and frustrating

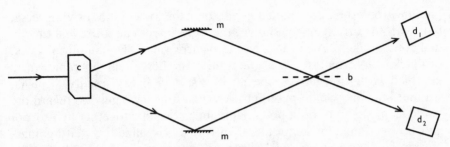

7.3 Multiple realities? Photons have uncanny abilities in this experiment per-formed at the University of California at Berkeley. A laser photon coming from the left is converted to identical-twin photons by the crystal c. After reflecting from mirrors m, the twins are brought together at a beam-splitter b, which transmits or reflects each photon with equal probability. Detectors d_1 and d_2 monitor the outcome. If the photons' paths have been of equal length, both go to the same detector—i.e., if one is transmitted by b, the other is always re-flected. The ghostly cooperation can be traced to the overlap or superposition of two alternative realities, which coexist here because the experimenter does not know which photon took which route.

results. One of these is the so-called quantum eraser dreamed up by physicist Marlan Scully, and designed to let an experimenter change his or her mind about what or what not to observe in a quantum system—even after the event![1] Fig. 7.3 shows one version of a quantum-eraser system, in which the quantum particles are photons from a laser. The first step is that a laser beam strikes a special crystal which converts each incoming photon into two weaker photons. These twin photons emerge from the crystal along different paths, but mirrors redirect them so they converge together again at a semitransparent plate called a "beam-splitter." The beam-splitter is a device to exploit the tunnel ef-fect: photons will tunnel through with a 50–50 probability. This means that the beam-splitter reflects half the light and transmits the other half. However, the geometry of the experiment is contrived so that both twins should arrive at the beam-splitter simultaneously. This entangles their destiny: although you cannot know in advance, because of quantum indeterminism, which photon will be transmitted and which reflected, the experimenters find that, *if* the lower photon is transmitted, then the upper one is *always* reflected, and vice versa. In either case, both pho-tons emerge from their encounter moving together along the same final path. Both paths—upper and lower in the diagram—are equally proba-ble. Detectors D_1 and D_2 lie in wait along each path to reveal the specific outcome in each individual case.

The reason both photons always end up in the same detector, upper or lower, hinges on the fact that, in the setup just described, the experi-menter cannot know which photon took which path. It may be that photon 1 took the upper path and photon 2 the lower path, or vice versa, but, with the arrangement shown, the experiment cannot reveal the actual routes taken. According to the bizarre rules of quantum physics, this lack of information about the routing implies a schizophrenic world in which *both* alternatives coexist in a sort of hybrid reality. That is, without knowing which photon took which route, we have to regard the world as made up of both potential realities existing together in a sort of ghostly overlap. This is not merely a way to visualize the weird goings-on, but leads to real physical effects. For example, we can tell that the two alternatives "photon 1 takes the upper path, photon 2 the lower path" and "photon 1 takes the lower path, photon 2 the upper path" *both* contribute to the outcome, because these phantom alternatives add together to produce results which are different from either alternative on its own—a process known as "quantum interference." In the present example, it is this interference of the alternative paths that produces the above-mentioned concordance, directing both photons to the same detector.

The interference arises as a consequence of the wave nature of light, and has to do with the fact that waves arriving in step reinforce, whereas those that arrive out of step cancel (I discussed this briefly in connection with the Michelson-Morley experiment in Chapter 2). Here the interference is between waves associated with one alternative reality combining with waves associated with the other alternative. The overlapping of these alternative-world waves can be convincingly demonstrated by slowly increasing the length of one of the paths until the waves associated with the alternative realities now arrive exactly *out* of step. In this case, interference causes *cancellation* of the wave, meaning that the two photons now go to *different* detectors—i.e., the detectors fire simultaneously. A slight additional increase in the path length brings the waves back into phase, and the photons go to the *same* detector once more. By gradually extending one path in this way, the experimenters can obtain a series of peaks and troughs—characteristic of an interference pattern —for the simultaneous firings of the two detectors.

The claim that a ghostly conspiracy of alternative half-realities leads to the cooperation of the two photons can be confirmed by modifying the experiment so that the individual photons are tagged in some way to determine their actual routes. This can be achieved by installing a simple gadget in the lower path that delivers a ninety-degree twist to the photon's polarization (Fig. 7.4). As a result, the photon taking the lower route will be identifiable, enabling the experimenter to tell which photon went which way. When this is done, the photon twins no longer invariably end up in the same detector, but instead behave independently and can trigger *both* detectors simultaneously even when the path lengths are the same. This experiment provides a clear example of wave-particle

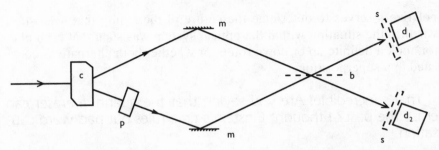

7.4 Changing the past? The setup shown in Fig. 7.3 is modified by introducing a device p into the lower beam designed to tag the photon taking that route (by changing its polarization). This yields knowledge of which photon traveled which route. The cooperation described for Fig. 7.3 is then destroyed—the photons act independently and can trigger both detectors simultaneously. However, optional polarizes s inserted in front of the detectors can be used to erase the vital tags, after the event. When this is done, the original hybrid reality is restored, with both photons going to the same detector once more, even though, by the time the erasure is done, the photons have already passed through the optical system.

duality, which I discussed in Chapter 3. When routing information is absent, the laser light behaves like a wave, producing interference. When a modification is made to allow the routing to be determined, the interference disappears, and the light behaves as if made of particles, with each photon taking a particular up or down route.

Surprisingly, it is not necessary for the experimenter actually to go ahead and measure the photons' polarizations—i.e., to determine the paths they have taken—for the change in detector behavior to be observed. The mere *threat* of obtaining such information is enough to destroy the ghostly superposition of hybrid phantom realities. It is our *potential* knowledge of the quantum system, not our actual knowledge, that helps decide the outcome.

The eerie feature of the Scully experiment, which has been performed by a quantum-optics group in the University of California at Berkeley,[2] is that the threat of obtaining routing information can later be retracted. To accomplish this, additional polarization twisters are installed in front of the photon detectors (Fig. 7.4) in such a way as to render the original polarization directions indistinguishable (i.e., wipe out the information) and thus restore the indistinguishability of the photon paths. When this is done, the original situation is recovered, with wavelike interference being observed once more. The mind-boggling feature of this experiment is that the retraction occurs *after* the photons have traversed the optical system! It is as if the photons somehow "know" in advance that the additional information-erasing polarizers lie in wait, and adjust their behavior accordingly. In effect, the decision to interpose the additional polarizers serves to determine the nature of the reality that *was*—i.e., whether the situation within the optical system was such that each photon took a definite up or down route, or whether both alternatives coexisted in a superposition.

That's incredible! Are you saying that the quantum eraser can erase the past? I thought Einstein's time rules out backward causation.

It's true that such experiments confirm Einstein's worst fears. But, although the actions of the experimenter can help decide the nature of the quantum reality in the past, the experiment cannot be used actually to send information into the past, which is the crucial point about causality. It is in fact the inherent uncertainty of quantum mechanics (which Einstein hated and never believed in) that miraculously comes to the rescue of Einstein's time. Because the experimenter doesn't know in advance which detector will be triggered (but only the 50-50 betting odds), she or he has no control over the photon-by-photon details. Any attempt to encode a message to be sent backwards in time would degenerate into white noise.

You still make it seem as if a human being can help shape the reality of the past. What would happen if the experiment was completely automated?

The Berkeley group has suggested this. Their idea is to replace each polarization eraser with another sort of beam-splitter that directs photons with different polarizations to different upper and lower subdetectors. Then scrutiny of which photons go to which subdetectors when, will automatically yield information about the routes the respective photons took. On the other hand, if the data from both sets of subdetectors are amalgamated together, this information will be masked. The output from the detectors can be stored in a computer, and the data analysis done at leisure at a later stage. The scientist can then choose, at whim, whether to inspect the amalgamated data with the which-route information safely camouflaged, or else to unpick the data to see which photons took which route on a case-by-case basis. Quantum theory makes a definite prediction about the outcome. The amalgamated data should show no rise-and-fall interference pattern in the simultaneous firings of the two detectors as one path length is extended, but if the experimenter chooses to separate out the data referring to the individual subdetectors, then an interference pattern should be apparent. In other words, the interference pattern characteristic of photons "taking both paths" is hidden in the total data describing the photons taking one path. The experimenter has the choice, when inspecting the computer data long after the experiment has ceased, whether to "look and see" which path the photons took, or to ignore this information and "observe" (i.e., reconstruct) a world in which both paths contributed.

This is all very confusing. Precisely when does each photon decide whether to take one path or, so to speak, both paths? When it passes through the polarization twister (or not), when it arrives at the beam-splitter, when it encounters the polarization scramblers, or the detectors, or when somebody decides how to arrange the data in the computer?

Your question doesn't have an answer. The commonsense idea that there is an objective reality "out there all the time" is a fallacy. When reality and knowledge are entangled, the question of *when* something *becomes* real cannot be answered in a straightforward manner.

But surely the act of twisting a photon's polarization has something to do with making it "decide" which of the alternative realities it will opt for?

Not so! It turns out it is not necessary actually to intervene in the photon's career to determine its routing. Incredibly, it is possible to obtain information about the path of a photon without doing anything directly to the photon of interest.

SPOOKY SIGNALS AND PSYCHIC PARTICLES

> We all *know* what light is; but it is not easy to *tell* what it is.
>
> DR. SAMUEL JOHNSON

Rather than tagging photons by a polarization twist, as in the arrangement described above, a different strategy was used in an experiment recently performed at the University of Rochester.[3] In this case, the laser light was *first* passed through a semitransparent mirror, splitting the beam in two, and then *each* beamlet was passed through a crystal converter to produce twinned photon pairs (see Fig. 7.5). Only one photon at a time enters the apparatus. The two upper light paths emerging from the crystals were allowed to intersect at a second beam-splitter, after the fashion of the Berkeley experiment, so that interference effects

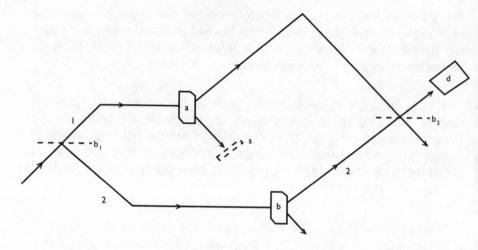

7.5 *Psychic photons. Beam-splitter b_1 divides a light beam into two paths, 1 and 2, which intersect again at a second beam-splitter, b_2. Any interference pattern at b_2 can be monitored by detector d. Each path contains a crystal to convert the incoming photon into a signal photon, which carries on to b_2, and an idler photon (i_1 or i_2), which can be directed to another part of the lab. If the two idler paths are merged, the experimenter cannot know which route the signal photon took: an interference pattern results. If the top idler path is blocked by an optional screen s, the experimenter can deduce the path of the signal photon, and the pattern disappears. Somehow the signal photons "know" what is happening to the faraway idlers.*

could be monitored by a photon detector. Photons ending up here were called "signal photons." Those emerging from the crystals along the two lower paths were called "idler photons." The idea of this arrangement is that, by observing the idlers, we can glean information about the routing of the signal photons. The incoming photon converts into a pair: a signal photon and an idler. If the idler is seen coming out of crystal A, then the experimenter knows that the signal photon is taking route 1. If the idler emerges from crystal B, then route 2 has been chosen by the signal photon.

So far there is nothing surprising involved. If the system is run in this manner, there will be no interference effects, because the experimenter is able to tell in each case which route the signal photon takes. As a result, the particle rather than the wave nature of the light is manifested. The novel step comes from *merging* the two lower paths in such a way that it would be impossible for the experimenter to determine from which crystal the idler photon had come. When this was done by the Rochester group, the signal photons created the distinctive interference pattern at the detector. Once again, the pattern arises because of the overlap of phantom alternatives. The worlds of path 1 and path 2 are superimposed on each other to form a hybrid reality. If the experimenters chose, they could unmerge the idler beams (e.g., by simply blocking the idler path from crystal A). When they did this, the behavior of the signal photons was dramatically altered: the interference pattern disappeared. The change occurred even though the idlers and the signal photons remained physically well separated at all times! So, without actually *doing* anything directly to the signal photons—merely by interrogating their idle twins in another part of the lab—the experimenters found that the signal photons adjusted their behavior obligingly. It's almost as if the signal photons are psychic: they "know" their twins have been questioned, and forced to divulge the details of the routing.

But how do they find out? Could it be that the act of observing the idler photons sends some sort of message across the lab saying: "Change your behavior, O twin! Your route has been revealed"?

Einstein considered the problem of secret quantum messages. He was long ago aware that quantum physics gets pretty threatening when "nonlocal" observations are involved—i.e., when *simultaneous* observations are performed at different locations in space. In 1935, now ensconced in Princeton and, in his mid-fifties, approaching the end of his productive career, he conceived of another thought experiment with colleagues Nathan Rosen and Boris Podolsky, known as the "EPR experiment" after its creators. The basic idea is that two quantum particles fly apart from a common point of origin, and observations are performed simultaneously on both particles when they are well separated. According to quantum mechanics, the state of the widely separated particles remains entangled in a manner that is impossible to square with the sort of commonsense reality Einstein yearned for. He wanted to believe that quantum particles such as photons are really "out there" with a complete set of well-defined properties (such as position, routing and polarization) before anybody decides to take a look at them. But it can be proved that, if Einstein's view were correct, the particles can comply with the rules of quantum mechanics only if they do somehow secretly communicate with each other across space (Einstein called it "spooky action at a distance").

But Einstein rejected the idea of hidden signals, because that implied an *instantaneous* dialogue between the separated particles. Apart from being absurdly conspiratorial (imagine two photons meters or more apart cooperating in the way they behave in their respective measuring apparatuses), it flagrantly contradicted the theory of relativity, which forbids faster-than-light signaling. If we recall the adventures of Ms. Bright

(p. 80), such signaling would imply the possibility of backward-in-time causation. So EPR instantaneous signaling, and the phenomenon of erasing the past that arises in the Berkeley and Rochester experiments, are really part and parcel of the same conundrum.

The latest quantum optics experiments are enough to make Einstein turn in his grave. But even worse could be to come. Recall that in the Berkeley experiment shown in Fig. 7.4, the decision of the experimenter either to observe wavelike interference patterns or not can be delayed until *after* the photons have traversed the optical system. That is disturbing enough. But the Berkeley group go one step further and claim that this crucial decision, which helps frame the nature of past reality, can even be delayed until after the signal photons have been *detected!*

To argue their case, Raymond Chiao and his colleagues at Berkeley have proposed a refinement of the set-up shown in Fig. 7.5. They point out that if the idler photons emerging along path i_1 are tagged in some way (such as by twisting the polarization), then the experimenter can tell which path each signal photon took simply by measuring the polarization direction of its associated idler. Once again, however, this information can be erased by inserting a second polarization twister further downstream. (Specifically, the first polarizer gives a 90° twist to the polarization of photons in the beam i_1, and the second polarizer is placed at 45° to the first.) But in each case the decision of whether or not to perform the second twist can in principle be left until *after* the signal photon has been detected at D. If the second polarization twister is left out, the signal photons behave like particles, but if it is inserted, then wavelike interference is predicted to reappear—in the form of distinctively patterned correlations between the signal photons and the idler photons.

Plans are in progress to perform experiments of this sort. They would not enable the experimenters to actually signal or alter the past. Rather they would demonstrate that the specific nature of past reality which is disclosed through the experimental observations isn't finally fixed until the entire experiment is brought to a close. Even when the signal photons have been detected, the record of the past remains not only incomplete but *undecided* because of the subtle long-range entanglement between signal and idler photons.

Einstein used the unacceptable nature of backward-in-time signaling as an argument to reject quantum mechanics, but Bohr retorted that it was Einstein's naïve view of reality that should be rejected. The particles simply do not have well-defined attributes prior to being observed, he said. In the event, Einstein's thought experiment has now metamorphosed into a series of real experiments, and the results have confirmed that Bohr was definitely right and Einstein sadly wrong. While strict backward-causation is smeared away by quantum fuzziness, a worrisome vestige of "ghostly" action survives in the results of Scully and company.

These experiments highlight the extremely peculiar nature of time in quantum physics. Although complying with the strict letter of the law governing Einstein's time, they violate the spirit of relativity by entangling actions in the present with the reality of the past.

FASTER THAN LIGHT?

Notwithstanding Solomon, in a race speed must win.

BENJAMIN DISRAELI

The conclusion of the various two-photon experiments, many of which have been performed only recently, is that it is not possible in general to say when things "actually happen" in quantum physics.

So there is no hope of telling how long a particle takes to tunnel through a barrier?

Strangely enough, it may still be possible. You see, there is a subtle difference between determining *when* a particle tunneled, and *how long* it took. If we are only interested in the total duration between start and finish, and not the actual moment of tunneling, there is a chance we can still measure it.

In fact, the Berkeley researchers attempted to do just that.[4] They based their experiment on the setup shown in Fig. 7.3; recall that if both photon paths are the same the photons will arrive at the beam-splitter simultaneously and, for reasons of quantum interference, go to the same detector. The optical arrangement in effect provides a racetrack to compare the travel times of the two photons.

Now suppose a barrier is inserted in one of the paths. Because the photon on that route has to tunnel through the barrier, it may not arrive at the meeting point at the same moment as its twin, in which case the delicate interference arrangements are upset, and there is a chance that one photon will go to each detector. However, by adjusting the length of the other route (the one that the twin took) to compensate, you can restore the situation and arrange for simultaneous arrival, and infallible cooperation in detector choice, once more. If the photon is slightly delayed by going through the barrier, then the twin's path will need to be slightly lengthened to compensate. By measuring the extra length, you can work out how long the photon took to tunnel.

When the experiment was actually performed, the results were amazing. With the barrier inserted, the photon that tunneled arrived first! In other words, the barrier seemed to speed the photon up. But the photon was already traveling at the speed of light, so on the face of it the photon that tunneled did so faster than light! The Berkeley group inferred a boost to the photon's velocity of some 70 percent—i.e., the photon tunneled at over five hundred thousand kilometers per second.

Did the Berkeley group cleverly create tachyons? Not really. Again, we have to be very careful about inferring, in the crazy world of quantum physics, that something is actually the case "out there" based on the results of specific experimental arrangements. Causality-busting tachyons require that we can exercise control over the moment of transmission and detection of the particles involved. It is not the same as deducing, *after the event,* that something might have exceeded the speed of light in the past, if we also know that, had we subjected that fleeting superluminal motion to observational scrutiny, we would not have been able to "catch it in the act" at a specific moment of time as told on a real clock.

THE TIME VANISHES!

Should we be prepared to see some day a new structure for the foundations of physics that does away with time? . . . Yes, because "time" is in trouble.

JOHN WHEELER

Clearly, the topic of time in quantum physics is a decidedly murky one, and for good reason. First, as we have seen, there is no such thing as a perfect clock in quantum physics—all physical clocks are themselves subject to quantum uncertainty. This smears their smooth running in an unpredictable manner, and might even cause them to run backwards. Second, Einstein's time is not Newton's time; it is flexitime, its malleability inseparably interwoven with the affairs of matter and gravitation.

Because the strange rules of quantum physics are supposed to govern all things, gravitational fields included, then not just clocks but *time itself* will be subject to quantum fuzziness. This brings us to the tough topic of quantum gravity. When quantum physics is applied to the electromagnetic field, you get photons and all the weird and wonderful phenomena that I have discussed above. In the case of the gravitational field, Einstein showed how it may be viewed as a warping or curvature of spacetime. Therefore, when quantum physics is applied to gravity, space and time adopt weird quantum properties too. This considerably exacerbates "the problem of time" in quantum physics, and a host of puzzles remain on the physicist's agenda to solve.

The core difficulty with quantum time harks back to the very notion of Einstein's time: there is no absolute and universal time. My time and your time are likely to be different, and neither is "right" or "wrong"; they are equally acceptable. Viewed in terms of four-dimensional space-time, different choices of time correspond to different ways of slicing up, or decomposing, spacetime into sections (recall Fig. 2.2 on page 74). Christopher Isham, Britain's top quantum-gravity expert, explains it thus:

> A central feature of the general theory of relativity is that all such decompositions of the space-time are deemed to be admissible and of equal status. In this sense "time" is a convention; any choice will do provided only that events can be uniquely ordered by the assigned values of time.[5]

The absence of any absolute, underlying time implies that physical processes cannot ever depend explicitly on time as such—for whose time will they choose? You may smell a paradox looming here—it seems to suggest that nothing can ever change in a quantum universe—but this is not so. The point is, rather, that the only meaningful way to measure physical change in Einstein's universe is to forget time "as such" and gauge change solely by the readings of real, physical clocks, not by some nonexistent notion of "time itself."

It must be said that many leading physicists are profoundly unhappy with the foregoing conclusion, and have labored hard to unearth some "true," intrinsic time buried obscurely within the mathematics of general relativity. They hope that some ingenious and subtle combination of quantities describing the geometry of spacetime may be revealed to possess the qualities one might expect for a universal measure of time, and that henceforth this universal time might serve as a genuine "background" for the measurement of change. So far, however, there is no evidence that such an intrinsic time exists.

But I thought you said earlier that there *is* a sort of universal cosmic time, and that Earth time almost coincides with it?

I did. However, cosmic time will not serve as an intrinsic time in any *fundamental* sense, because its existence depends, you will recall, on the fact that the state of the universe is highly uniform and symmetric on a large scale. A *general* spacetime will not possess this uniformity. According to convention, the job of the cosmologist is to take the laws of physics and explain the universe, not the other way around (as Milne believed). We want to regard the universe as a gigantic mechanical system subject to the laws of quantum physics, and, we hope, to account for why there is large-scale uniformity. But to carry out this central task of quantum cosmology, we need to explain how the universe evolves with time, without referring to time in any basic way!

Surely you can use the expansion of the universe itself as a clock?

You can. A *particular* slicing of spacetime into spatial sections *will* show how the geometry of space evolves with the time coordinate. In the simple case of a uniform universe, space just expands at a certain rate. But if we take a different method of slicing, we get a different description—e.g., a different size and expansion rate—at the same particular value of the time coordinate. The point about Einstein's time is that all such descriptions must be equivalent; the value of the time coordinate itself is arbitrary.

If you nevertheless go ahead and use an arbitrary time coordinate anyway, and treat the motion of the universe just like that of any other mechanical system, then you can use Einstein's gravitational-field equations to write down equations of motion for the universe, and identify familiar quantities such as the total energy. But here lies the rub. For the equations to remain valid whatever flexitime (i.e., slicing) you opt for, it turns out that the total energy of the universe is constrained to be exactly *zero*. So Einstein's view of time forces us to conclude that, if the universe as a whole is naïvely treated like a mundane mechanical system, its energy is obliged to vanish. This remarkable result, known to physicists for many years, has profound consequences for a quantum description. In quantum physics, energy always goes hand in hand with time. In a sense, the amount of energy determines the rate at which time passes—the beat of the quantum clock, if you like. No energy means the quantum clock ceases to tick: time bafflingly drops out of the physical description altogether. So quantum cosmology, treated in this manner, makes no reference at all to time: in effect, time has totally vanished too! Spacetime, the very entity on which Einstein's theory of relativity was founded, has been substituted for a motley collection of spaces of different geometry, but with no time left to stitch them together. Like the dog that failed to bark in the Sherlock Holmes story, the cosmic clock that failed to tick seems to be a crucial clue that may help us to solve the riddle of time, but we lack the sleuth's fabled power of reasoning to crack the problem.

That's all pretty mysterious, don't you think? What has happened to time? You make it sound as though it never really existed.

It has evaporated in a puff of quantum fuzziness, in the same way that other precise notions, like position and trajectory of motion for particles, disappear in conventional quantum mechanics. Quantum cosmology has abolished time as surely as the mystic's altered state of consciousness. For a typical quantum state in this theory, *time is simply meaningless*.

So where did time come from? If it has no fundamental physical existence—if it wasn't, so to speak, made in the big bang—what produced it?

A valid point. I am the first to concede that in the everyday world time does have an important significance. No theory of the universe can be credible unless it allows for some sort of notion of time to emerge from the quantum fuzz. What the latest thinking among quantum cosmologists suggests is that time is merely an *approximate* and *derivative* notion. Calculations have eagerly been done in an attempt to elucidate precisely how cosmic temporality "congeals" or "crystallizes" out of the timeless quantum frolic of contorted geometries that clothed the big bang. As I write, these calculations remain (to my mind) as slippery as the time they seek to capture. All that seems clear is that a general quantum state of the universe has no well-defined time at all.

The difficulty these calculations confront is that the quantum fuzziness doesn't simply go away of its own accord. It afflicts not just the identities of space and time, but the *geometry* of spacetime too. In a quantum description, there is no single spacetime with a well-defined geometry that is "there"; instead, you must imagine all possible geometries—all possible spacetimes, space warps and timewarps—mixed together in a sort of cocktail, or "foam," after the fashion of the alternative realities represented by the photon paths that I discussed in the previous chapter. Somehow, from this foamy mess, some semblance of a space and time with a specific geometry has congealed. Nobody knows precisely how the congealing happened, but there is reason to believe it may demand a very special set of circumstances. That is, if you just take any old foaming quantum big bang, you won't end up with a well-defined time. The general rule is: once foamy and fuzzy, always foamy and fuzzy. Apparently, only *very special* initial conditions—that is, only universes which start out with very specially configured foam—will evolve into approximately "classical" (i.e., nonquantum) realities, possessing time, space and well-defined macroscopic material objects. For reasons we know not (though see my book *The Mind of God* for a go at them), the quantum state of our universe, fortunately, *is* one of those very special states that permit time to emerge from this primordial jumble, as the universe "evolves" away from the big bang, in a fuzzy and ill-defined way. And that is good news, because life in a universe without any sort of time would be difficult.

If these ideas are on the right track (and they are certainly highly speculative), then the quantity called "time" that is so crucial to our lives, and our description of the physical world, may turn out to be an entirely secondary concept, unrelated to the basic laws of the universe. History will have turned full-circle since Newton, who placed time at the center of his description of reality. Now we see that time may have originated almost by accident. We can imagine that, in the beginning, close to the big bang, time did not exist. Only because the quantum state of the universe is peculiar has time emerged in an approximate way—as a sort of relic—from the timeless primeval stirrings of the nascent cosmos. It may seem alarming that quantum physics abolishes time close to the big bang, but there is a welcome compensation: it may be precisely the loophole that is needed to explain how the universe came into existence in the first place.

CHAPTER 8
IMAGINARY TIME

So maybe what we call imaginary time is really more basic,
and what we call real time is just an idea that we invent to
help us describe what we think the universe is like.

STEPHEN HAWKING

Because *Mathematicians* frequently make use of Time, they
ought to have a distinct Idea of the meaning of that Word,
otherwise they are Quacks. . . .

ISAAC BARROW

THE TWO CULTURES REVISITED

Professors . . . are wholly useless, most of them.

JOHN MAJOR, BRITISH PRIME MINISTER

I'll never forget the moment I first saw—or, rather, heard—Stephen
Hawking. The year was 1969, and I was attending a one-day conference
on gravitation theory at King's College in London, located in the Strand,
close to the famous Fleet Street. It was a day's light relief from the
rigors of my thesis work. The speaker at the time was the world-famous
mathematician Roger Penrose. He was in the middle of his lecture when
he was suddenly interrupted by a voice from the front row. My initial
impression was that a drunk or a madman had sneaked into the audito-

rium, intent on mischief (such things are not unknown at physics conferences). The interjection, delivered in a thick drawl that was completely
incomprehensible to me, continued for a full two minutes. To my astonishment, Penrose just stood there patiently for the duration, and then
proceeded to articulate a lengthy technical answer to what had evidently
been a very pertinent question from the young Hawking.

I never fully mastered Hawking's slurred speech, though I learned to
follow the drift of his remarks. In the days before he was equipped
with a voice synthesizer, conversations with him were always open to
misunderstandings, sometimes of a humorous character. On one memorable occasion in Boston, while we were discussing the program of a
conference over a meal in a restaurant, Stephen repeatedly asked
whether we were going to have any wine. After unsuccessfully trying to
get him to select a bottle from the wine list, I suddenly realized he was
talking about Weinberg, the physicist.

Even in those early days, Hawking was interested in the problem of
whether time had an origin, or extended backwards forever. Did the
great cosmic clock have a first tick, as it were, or has it been ticking
away for all eternity? When, twenty years on, Hawking encapsulated his
deliberations on the subject in the book *A Brief History of Time*, he
became an instant celebrity, widely compared to Einstein. Perhaps inevitably, his high public profile provoked a backlash. Especially outraged
by Hawking and his daring ideas were the British chattering classes. By
tradition, British intellectual life is dominated by the arts-and-literary
fraternity, as readers of C. P. Snow will be aware. Indeed, scientists are
rarely even afforded the status of "intellectuals." Science, to the extent
that it is considered at all by British opinion-formers, is regarded as at
best a necessary evil required to propel money-spinning technology, and
at worst a technocratic conspiracy. There is a tacit assumption that
scientific theories are part of a gigantic confidence trick designed to
inflate the power and importance of self-serving scientists. Scientific
discourse is treated with suspicion as an esoteric code, created to maintain the exclusivity of the club and deliberately designed to dazzle nonscientists with impenetrable mathematics and jargon.

So long as scientists restrict themselves to their laboratories, they are
tolerated by the literary establishment—shrugged aside as nerds of little
consequence—and the implications of their obscure and incomprehensible work are ignored. But what incenses these opinionated literati most
is when scientists dare to tangle with "meaning-of-life" issues. The arts-
and-literary community have long believed they possess a God-given
monopoly on such matters. That Stephen Hawking, a scientist, had the
audacity to draw profound conclusions about the origin of the universe,
the role of a Creator, or the place of mankind in the cosmos from his
arcane discipline was considered beyond the pale. Though I do not agree

with all of Hawking's conclusions, he merely articulates what is a fairly common position among scientists, and he should not be vilified for doing so.

The chorus of anger that rose in response to Hawking's book took the form of public denunciations by self-righteous politicians and journalists, and near-hysterical diatribes in the British press by renowned writers and academics. Their discomfort was fueled by the fact that few of these people could understand the content of the book, as virtually none of them had any scientific education and most were generally hostile to science on ideological grounds anyway. The feeble argument was trotted out that any important truth ought to be transparent to all thinking people. It was a case of "I am well educated"—i.e., well versed in the arts and literature—"and I can't understand these claims by physicists and cosmologists. Therefore, the claims must be nonsense and the scientists are frauds."

To make this point, commentators adopted the habit of taunting scientists by asking the question: "What happened before the big bang?" The feeling seemed to be: "You scientists think you're so clever at explaining everything. Well, even if you explain the big bang, you still haven't explained what was there before, have you?"

HOW TIME GOT STARTED

Unfortunately, the aforementioned challenge betrays an ignorance not just of science but of the history of philosophy and theology too. As we have seen, Augustine long ago proposed that the world was made *with* time and not *in* time. He recognized that time is itself part of the physical universe—part of creation—and so any talk about "before" creation is meaningless.

It's all very well, counters the skeptic (who rather sides with the British literati), claiming that Augustine sorted it all out. But baldly saying that time did not exist before the universe came into being is just a lot of words. How can we imagine such a thing? How can time suddenly start up on its own?

Augustine was more interested in theology than physics. As I have mentioned, his idea neatly solved the conundrum of what God was doing before he made the universe. Yet the problems of time and creation did not go away. Most theologians and scientists still assumed that time is everlasting and had no beginning. If, therefore, the universe of matter and energy had a definite origin (e.g., it was created by God at some specific moment), then there must have been an initial singular event, within time, at which the universe abruptly started to exist.

In the seventeenth century, Gottfried Leibniz, who wanted to believe that God created the universe a finite time ago, was nevertheless disturbed about why God, who is supposed to be perfect and immutable, suddenly decided to make a universe at some particular moment:

> For since God does nothing without reason and no reason can be given why He did not create the world sooner, it will follow either that He created nothing at all, or that He created the world before any assignable time, that is, that the world is eternal.[1]

The topic was taken up by Immanuel Kant, who craftily advanced arguments that cast doubt on *both* alternatives. The universe cannot be unlimited in past time, he reasoned, because that would mean that an infinite number of events or succession of states of the world must have occurred. But since infinity can never be attained "by successive synthesis," the hypothesis of an eternal universe has to be false. On the other hand, if the universe came into existence at some particular moment in time, then there must have been a time before it existed (Kant called this "void time"). But then he argued, obscurely, that nothing can originate in a void time, "because no part of any such time contains a distinctive condition of being, in preference to that of non-being." Kant accepted that to escape from his temporal dilemma would mean denying "the existence of an absolute time before the world," yet this he was not prepared to do, Augustine notwithstanding.[2]

What sense can be given to the idea of a time before the universe? If there are no "things"—just an eternal void where nothing happens—concepts like succession and duration don't seem to have any meaning at all.

Many people have an image of the epoch before the universe as a dark, inert, empty space. But for the modern cosmologist, neither time nor space existed before the big bang. The origin of the universe means the origin of space and time as well as matter and energy.

If time didn't always exist, then surely there must have been a discontinuity at which time abruptly "switched on"? And this means there would have been a first event—or First Event. The First Event can't be like other, ordinary events, because nothing came before it. It would be an event without a cause—a singular, supernatural event, surely?

Hawking's early work focused on the problem of the First Event. He was able to show, using the general theory of relativity, that the origin of the universe was indeed singular—in the rather precise mathematical sense I introduced in Chapter 4. If the simple model of the big bang is pursued to its ultimate limit, then the universe was infinitely compressed at the very beginning. This state has an infinite gravitational field, which represents an infinite warping of spacetime. You can no more continue spacetime beyond such a singularity than continue a cone beyond its apex.

So the first event was a spacetime singularity—a state of infinite density and curvature?

Not exactly. There is a subtlety here. The singularity (which is in any case a mathematical artifact) is defined to be a *boundary* to time, not strictly part of time itself—not actually an event as such. The singularity bounds time in the past, implying that time has not endured forever. Nevertheless, there need not have been a first moment.

What? Surely, if time has not endured forever, there must have been a first moment?

No. Is there a smallest number greater than zero? Clearly not. Try picking a number (one-billionth, one-trillionth . . .). That number can always be halved, and halved again, to obtain ever-smaller numbers. If time is continuous, then at no moment (one-billionth of a second, one-trillionth of a second . . .) would there have been *no* preceding moments. Of course, time may not be continuous. The great cosmic drama could be like a movie—a sequence of static frames, run past us at such a speed we don't notice the joins. It may just give the illusion of continuity. Theories involving "chronons"—atoms of time—have been proposed (most notably by David Finkelstein, whom I mentioned in Chapter 4), but without much success. On the experimental front, physicists routinely study sequences of events happening on a time scale of about a hundred-trillion-trillionth of a second, and no hint of temporal discontinuity has shown up yet. So, if there *are* chronons of time, they must be pretty brief.

Okay, I can see there are some technical mathematical quibbles. But, first moment or not, a singular origin to the universe means that time switched on suddenly, for no apparent reason. Such a "happening" (I won't call it an event) seems pretty supernatural. I don't see how the origin of time as such can be encompassed within the scope of science.

That was the general belief until a few years ago. The choice seemed simple enough: either the universe (and time) had no beginning, and have existed forever, or there was a singular beginning that couldn't be explained by science. Either way, there were problems. It all changed, however, when physicists started taking quantum effects into account. The crucial property of quantum physics is that cause and effect aren't rigidly linked, as they are in classical, commonsense physics. There is indeterminism, which means some events "just occur"—spontaneously, so to speak—without a prior cause in the normal meaning of the word. Suddenly physicists became aware of a way for time to "switch itself on"—spontaneously—without being "made to do it."

THE HARTLE-HAWKING THEORY

The notion of time fades gracefully away. . . .

CHRISTOPHER ISHAM

Stephen Hawking and James Hartle of the University of California at Santa Barbara sketched a way in which time could switch itself on, quantum-mechanically, at the big bang. They used a mathematical scheme that combined Einstein's time (and space) with the laws of quantum physics in a highly suggestive way. I should make it clear at the outset that the Hartle-Hawking theory is pure speculation based on somewhat shaky foundations, but at least it represents an honest attempt to tackle systematically what is perhaps the ultimate scientific challenge.

The cornerstone of their theory is something that Hawking has called "imaginary time." Unfortunately, a lot of people have taken this to mean something rather mystical, like "the time of our imagination." Others think it means a type of time which we can only imagine, not the "real" time of experience. In fact, the word "imaginary" is being used here in a technical mathematical sense, and has nothing whatever to do with imagination.

Let me explain. At school we are taught how to square numbers. For example, the square of 2 is $2 \times 2 = 4$, the square of 3 is $3 \times 3 = 9$, and so on. Going the other way is called "taking the square root." Thus the square root of 4 is 2, the square root of 9 is 3, etc. More advanced students learn how to square negative numbers. The rule is that multiplying two negative numbers gives a positive number, so $(-3) \times (-3) = 9$. This means there are *two* numbers which, when squared, give 9 (namely, 3 and -3). Conversely, if you want to take the square root of 9, the correct answer is 3 *or* -3.

A problem arises if you want to take the square root of a *negative* number, like -9. Because both negative and positive numbers, when squared, give a *positive* number, no ordinary number can be squared to give a negative number. If you want to talk about the square root of negative numbers, you have to invent some new numbers for the job, numbers which aren't contained in the familiar set 1, 2, 3 . . . or $-1, -2, -3$ . . . This was done in the sixteenth century. The new numbers were called "imaginary," not because they are any less real than "ordinary" numbers, but because they don't crop up in everyday arithmetic of the sort needed to count sheep and money. The suggestive term "imaginary" is typical mathematical jargon. There are irrational and transcendental numbers too, not to mention real, complex, rational and transfinite numbers and vulgar fractions. The names are just historical.

Because imaginary numbers are new, we cannot use any of the symbols reserved for "normal" numbers, so letters are used instead. Let's start with the simplest imaginary number, the square root of -1. It is denoted i. Thus $i \times i = -1$. This is simply a definition. Fortunately, it is not necessary to have an endless list of funny new symbols for imaginary numbers. Only one new symbol is needed, because all other imaginary numbers can be constructed by multiplying i by a "real" (i.e., ordinary) number. For example, the square root of -9 is $3i$, and so on. Imaginary numbers may seem unfamiliar, but they are widely used in science and engineering as well as in mathematics, often resulting in great simplification.

What has all this got to do with time? The connection goes all the way back to the work of Hermann Minkowski. Recall from Chapter 2 how Minkowski explained that a single "spacetime" continuum followed naturally from Einstein's special theory of relativity. Time was treated more or less as a fourth dimension by Minkowski, almost like space. But not quite. There *is* a difference in the way that time and space enter the description of spacetime. To see the difference, you have to look at the concept of distance in spacetime. Now, the distance between two points in space is well enough understood; it means the length of a ruler that connects the two points in a straight line. Distance in time is also easy; the interval between two events is simply the time difference told on a clock at rest in the reference frame of interest. But what do you do about mixing the two when space and time are merged in a unified spacetime?

Suppose you want to know the spacetime interval between New York at one o'clock and London at two o'clock? Minkowski gave the rule you need to compute this. First step: take the time difference and multiply it by the speed of light. This turns units of time into units of space. Thus one second becomes three hundred thousand kilometers (because light has a speed of three hundred thousand kilometers per second). Second step: *square* the result. Third step: square the distance in space (in kilometers). Fourth step: *subtract* the first number from the second. This is unusual. Normally when combining distances you use addition, but when time is involved you must subtract, a procedure that will turn out to hold the key to our concerns. Final step: take the square root. You now have the interval between two events in *spacetime*, expressed in kilometers.

Let me try an example. As the speed of light is so great, a little bit of time (e.g., one second) is worth an awful lot of space (three hundred thousand kilometers), so, to make the example interesting, I shall compute the spacetime distance between Earth at one o'clock and something far away—the sun—at 1:05 P.M. The Earth–sun distance is 150 million kilometers, so squaring gives 22,500 trillion square kilometers. Five minutes multiplied by the speed of light is 90 million kilometers, and squaring that gives 8,100 trillion square kilometers. We now have to do the crucial subtraction: 22,500 trillion minus 8,100 trillion equals 14,400 trillion. Finally, taking the square root, we get 120 million kilometers for the *spacetime* interval between these two events. Notice this is *less* than the spatial distance by 30 million kilometers.

Obviously, the bigger the time interval between the two events, the smaller the final answer will turn out to be. If we took the second event to occur at 1:08 P.M., we would compute a spacetime interval of only 42 million kilometers instead. With 8⅓ minutes' time difference, the spacetime interval would actually shrink to zero. This comes as a surprise. How can two events, separated in both space and time, have *zero* separation in spacetime? One way to see how is to note that the zero answer occurs in this example when the time difference is exactly equal to the time it takes for light to travel between Earth and the sun. Remember the twins, Ann and Betty? Betty's journey took less and less time, in her frame of reference, the nearer she got to the speed of light. At the speed of light itself, time stands still. The theory of relativity does not permit Betty to achieve this speed, but a light pulse can. From the point of view of the pulse, no time at all elapses as, in our frame of reference, it sweeps across the solar system. It is here, then it is there—instantly! Lightwise, there *isn't* any separation between Earth at 1:00 P.M. and the sun at 1:08⅓ P.M.

Trouble comes when the time difference is greater than 8⅓ minutes. Suppose we take 1:10 P.M. The time-squared number is now 32,400 trillion. This is *bigger* than the 22,500 trillion from which we have to subtract it. So the result is a *negative* number: −9,900 trillion. But now we confront the final step: take the square root to find the spacetime distance. Taking the square root of a negative number means we get an *imaginary* number for the answer. This is nothing to get too excited about. Physically, if the spacetime distance is imaginary, it simply means that the points have a greater separation in time than they do in space. The simplest example is for two successive events at the same place.

There is then *zero* spatial separation, so the answer has to be imaginary. For example, New York at 1:00 P.M. and New York at 1:05 P.M. are separated in spacetime by 90,000,000*i* kilometers.

The fact that *i* crops up when we compute some spacetime intervals but not others is a sign that space and time don't mix too intimately. The *i* flags time intervals, whereas the absence of *i* means we are dealing with spatial separations: there is a clear distinction. So, even though Einstein's space and Einstein's time are interwoven in Minkowski's spacetime, space is still space and time is still time. Time may be the fourth dimension, but it is not a *spatial* dimension, as those *i*'s tell us. It is because spacetime distances get small when we combine space and time separations near the speed of light that the geometry of Minkowski space has the wonky form I referred to in Chapter 2.

Now I get to the bit about imaginary time. If we multiply time intervals by *i*, then they are no longer imaginary numbers but real numbers, exactly like the spatial intervals. This is because *i* times an imaginary number gives a real number (remember, $i \times i = -1$). So, if we adopt the fiction that time intervals are imaginary numbers, then space and time become identical as far as the rules of Minkowski space are concerned, and time really is just a fourth dimension of space.

Of course, the world is not actually like that, but Hawking's idea is that it may have been so once. (In the quote that heads this chapter, Hawking hints at a belief that maybe the world is actually like that even today. I have to disagree.) Specifically, time might have been imaginary (i.e., just like space) near the big bang. This idea wasn't plucked out of midair. Imaginary quantities pop up all over the place in quantum physics, and sometimes cause mathematical difficulties. For years, physicists have occasionally cheated by artificially treating time as if it were imaginary so that they can finish their calculations instead of remaining stuck. Sometimes this is just a dubious device which may nevertheless give the right answer; sometimes it can be justified by a deeper theory. In the case of applying quantum physics to cosmology, the project Hartle and Hawking embarked upon, there is indeed some justification. As I keep remarking, quantum uncertainty has the effect of smearing out, or introducing a fuzziness into, all measurable quantities on a microscopic scale. This includes space and time. When they are stitched together as spacetime, it can happen that the quantum fuzziness smears together some space bits and some time bits. In other words, time intervals might be indistinguishable from space intervals: hence "imaginary time." The smearing and conflating of space and time isn't noticeable in daily life of course. It is limited to tiny intervals (about 10^{-33} centimeters of space and 10^{-43} seconds of time). Nevertheless, if it occurs, it dramatically alters the nature of the First Event problem.

The quantum smearing isn't a discontinuous thing: time can be a little bit smeared or a lot smeared, which means it might be a little bit spatialized or a lot spatialized. We can imagine a continuous sequence where time "starts out" as space and gradually "turns into" time. (Or, in reversed-time language, time gradually fades away as one goes back in time towards the origin.) This statement abuses language in several ways. Time is always time; it doesn't really "turn into" anything. More accurately, what we call time may have once had some of the properties we normally associate with space. And "gradually" means over a mere 10^{-43} seconds, which is pretty fast by most standards! Nevertheless, in this theory there is no singular origin of time, no abrupt "switching on" at $t = 0$.

On the other hand, time does *not* extend back infinitely. It is certainly bounded by the big bang in the Hartle-Hawking theory, as it is in the conventional theory, where there is a spacetime singularity blocking the physical universe off that way. Many people mistakenly suppose that Hawking has done away with the origin of the universe. This is quite wrong. In his theory, time is definitely bounded in duration, but there is neither a First Event nor a sudden, singular and supernatural origin. And apart from the above funny business around 10^{-43} seconds, the aftermath of the big bang is much the same as before.

These ideas can be applied to the end of the universe too. We can imagine that time may not go on forever, but will continuously fade away, spatializing itself, in the same manner as it originated. There would then be no Last Event, no Final Moment, no Judgmental Nanosecond of reckoning. But the future would be bounded too.

You must understand that the description I have given of Hartle and Hawking's work covers a multitude of sins. In particular, the simple statement that time "emerges" continuously from a dimension of space is easy enough to state in words, but the mechanism of that emergence is far from clear. In fact, as I remarked in the previous chapter, it is still a deep mystery precisely how a well-defined notion of time (and space) emerged from the quantum fuzziness of the big bang.

IMAGINARY CLOCKS

It's all very well, complains the skeptic, for you to float abstract mathematical concepts like imaginary time and fuzzy quantum time, but what connection, if any, have these theoretical times got with "real," honest, everyday time—human time, if you like? How do you even measure imaginary time anyway?

When physicists and cosmologists use the word "time" in connection with the very early universe, they are abstracting and extrapolating and idealizing in several ways. First, no known clock can measure intervals of time less than about a trillion-trillionth of a second. Clocks of greater temporal resolution may exist, but we haven't yet discovered them. To discuss durations less than this (admittedly brief) interval, you have to assume (a) that time really is continuous on shorter scales, and (b) there are at least some periodic physical processes that happen faster than this which can be used to define a clock. Moreover, your clock had better be small. If no physical effect can travel faster than light, then a clock cannot measure time more accurately than the time it takes light to travel between the components of that clock. In a trillion-trillionth of a second, light travels a distance less than the size of an atomic nucleus. So the clock would have to be some sort of subnuclear entity, such as a particle. Then there are all those tricky problems about measuring quantum time with quantum clocks that I discussed in Chapter 7.

Even supposing that an appropriate clock can be (approximately) defined to chronicle hypothetically the early history of the universe, you have to assume that it remains at rest in the privileged reference frame in which cosmic time is measured. This is certainly a fiction. If you imagine the experiences of, say, a subatomic particle in the hot, dense early universe, it will suffer countless high-speed impacts, and will spend most of its time traveling at close to the speed of light, being knocked every which way. In the reference frame of the particle, there will be a huge time-dilation effect, which means that the time which our "clock" experiences will be a vastly extended measure of "real" cosmic time.

Rather than imagining some hypothetical and very artificial sort of clock being used to gauge the activity of the primeval cosmos, we could turn the problem on its head and use the *activity* to gauge the *time*. In other words, we could define the unit of time to be the duration required for some sort of physical activity to occur. For example, the average duration between collisions of particles is a possible measure of activity. Because the universe gets hotter and particle motions become more frenetic as the origin is approached, this measure of time will stretch at an escalating rate as we go back in time; it may even be infinite, depending on how things really were near the origin. If that were so, then in a certain sense (from the standpoint of the particle) the universe would have existed forever!

But how do these hypothetical clocks relate to human time— time as experienced by us?

Well, as there were no humans around just after the big bang, this question is a bit ambiguous. You can ask what sort of time humans experience *now*, and find a clock that more or less measures this contemporary human time (e.g., the clock on my wall); then you can imagine a hypothetical clock that agrees with this clock but somehow magically could have survived unscathed the extreme conditions of the big bang. As far as this time scale is concerned, the universe apparently began several billion years ago.

However, you could view things a bit differently. What is it, after all, that determines the measure of human time? This is a major topic which I shall take up in Chapter 12, but for now let me just make the point that our perception of time clearly has something to do with our brain processes. If our brains worked at twice their actual speed, then one second would seem the way two seconds does to us now.

The speed of physical processes depends on temperature: the hotter the system is, the faster things—and by implication thoughts—will happen. Of course, the temperature of the human brain is regulated within very precise limits, but we can imagine sentient beings living at higher temperatures with a higher metabolic rate experiencing a more rapid subjective time compared with us. If (somehow!) the frenetic processes that occurred in the early universe supported some sort of mental activity, then the subjective time of this primeval being would rise towards infinity as the origin of the universe was approached. In other words, if you measure the time that has elapsed since the big bang in terms of a hypothetical *sentient* time (the primeval being's sentience, not ours), it is probably infinite.

The same state of affairs would prevail if the universe were to collapse to a big crunch. As it neared the end, so the universe would become hotter and hotter and the rate of activity would escalate, perhaps without limit. The physicist Frank Tipler has speculated (wildly) about a superbrain in the far future that spreads across space until it encompasses the entire universe.[3] This cosmic individual might be capable of thinking an infinite number of different thoughts before the final crunch came, by harnessing the escalating physical activity to use as its own brain processes. As the crunch approached, so it would think faster and faster. Whether the speed of the superbrain's thoughts could literally grow without bounds depends crucially (as explained in Chapter 5) on the limitations of fundamental physics, such as the speed of light and quantum smearing. Tipler has investigated this. He finds that for most collapsing-universe models, such limitations stymie the superbrain's quest for unbounded thinking power. But there do exist certain complicated models at which known limitations may be evaded. Tipler claims the superbeing would then literally live forever, in the sense that its *subjective* time would be infinite, even though, on an extrapolation of human time, its existence is bounded in the future.

The converse scenario has been discussed by Freeman Dyson, who envisages a different class of sentient beings eking out a sad and desperate existence trillions of years down the track, when the universe has grown ice-cold and is approaching a heat death.[4] If the universe does not collapse, as Tipler supposes, but goes on expanding forever, then all known sources of energy will dwindle towards zero, and our surviving descendants would inexorably slow up, both physically and mentally. They would be compelled to harvest scraps of energy for fuel from ever-larger cosmic domains, and conserve it through long periods of hibernation to make it last. In this case, the subjective time of these hard-pressed and slow-witted individuals would become more and more stretched relative to cosmic time. (You can learn more of the adventures of Tipler's seared superbeing and Dyson's frigid fossickers from my book *The Last Three Minutes*.)

Tipler and Dyson offer sharply contrasting images of the end of time: one in which mental activity speeds up and time slows down, the other in which mental activity slows down and time speeds up. But there is a third possibility too. Time might go into reverse.

CHAPTER 9
THE ARROW OF TIME

All change, and time's arrow, point in the direction of
corruption. The experience of time is the gearing of the
electrochemical processes in our brains to this purposeless drift
into chaos as we sink into equilibrium and the grave.

PETER ATKINS

CATCHING THE WAVE

One of the most interesting and offbeat scientists of the postwar years
was David Bohm, an American-born theoretical physicist who worked
mainly in London, at Birkbeck College. I first met Bohm when I was
twenty-three, an eager and inquisitive Ph.D. student at London University. The purpose of the meeting was to discuss an aspect of my thesis
project which concerned a nagging paradox about the nature of time.
Roughly stated, the paradox is as follows. We take it for granted that,
when a radio station transmits a signal, we receive the signal on our
radio set at home *after* it was sent out by the transmitter. The delay
is not very long—just a fraction of a second from point to point on
Earth—so we are normally unaware of it. But a telephone conversation
relayed via satellite can introduce a noticeable time lag. Anyway, the
point is that we never get to hear the radio signal *before* it is sent.

Why should we? you may ask. After all, effects don't normally happen
before their causes. The problem that lay at the root of my worries stems
from the mid-nineteenth century, when James Clerk Maxwell wrote
down his famous equations that describe the propagation of electromagnetic waves such as light and radio. He did this while working at King's

College in London, just a mile or two from Birkbeck. Maxwell's theory predicts that radio waves travel through empty space at the speed of light. What Maxwell's equations do not tell us, however, is whether these waves arrive before or after they are transmitted. They are indifferent to the distinction between past and future. According to the equations, it is perfectly permissible for the radio waves to go backwards in time as well as forwards in time. Given a pattern of electromagnetic activity, such as that corresponding to radio waves from a transmitter spreading out through space, the time-reversed pattern (in this case, converging waves) is equally permitted by the laws of electromagnetism.

In physics jargon, forward-in-time waves are called "retarded" (as they arrive late) and backward-in-time waves are called "advanced" (as they arrive early). Because we don't seem to notice advanced radio waves, or advanced electromagnetic waves of any sort, the advanced solutions of Maxwell's equations are usually simply discarded as "unphysical." But what justification have we for doing that? Is there another law of physics, in addition to the laws of wave motion, that commands: "No advanced solutions in *this* universe!"? If not, what else might lead nature to prefer retarded waves over advanced waves, given that both varieties apparently comply with her laws of electromagnetism?

I had been electrified by this puzzle ever since attending a mind-stretching meeting of the Royal Society in 1967, at which the Cambridge astronomer Fred Hoyle presented his own solution to the time-asymmetry enigma. Hoyle was convinced that the answer lay with the way the universe expands. The suggestion that what happens in my radio is somehow linked to the destiny of the cosmos I found positively enthralling, and I resolved to work on it myself. I focused my investigations on the simplest system that could transmit and receive electromagnetic waves—a single atom. If an electromagnetic wave falls on an atom in its normal, or ground, state, the atom can be induced to make a quantum leap to an excited state by absorbing a photon from the electromagnetic radiation. This corresponds to an antenna "receiving" the wave. Conversely, if an atom is initially in an excited state, it can make a quantum transition to its ground state and emit a photon. This corresponds to transmission. At the quantum level, then, the process looks pretty symmetric: the time reverse of an atom absorbing a photon is an atom emitting a photon.

In fact, Einstein had long ago appealed to this very symmetry between emission and absorption of photons to compute the rate at which an excited atom would spontaneously emit a photon into free space. He did this in 1916, shortly after the collapse of his marriage to Mileva, and well before quantum mechanics had been fully worked out. His insightful calculation also includes an expression for the rate at which an atom will radiate a photon if it is being bombarded by other photons, a process called "stimulated emission," which was to become the principle behind the laser developed nearly half a century later.

The symmetry between atomic emission and absorption of photons rests, however, on a hidden assumption. When, following Einstein, you compute the rate at which an unexcited atom will absorb photons, the textbooks tell you to assume that all the photons coming at the atom are uncorrelated. In wave language, this means that all the individual electromagnetic waves corresponding to those photons are completely scrambled up: their phases are distributed at random. I wanted to know where this random phase assumption, so crucial to establishing the time symmetry between emission and absorption of photons by atoms, came from. So I went to see Bohm.

Although he was to became a cult figure with a worldwide following, Bohm was a rather retiring man. He was inclined to become agitated as he warmed to his subject, though. His vocabulary was good, but in his excitement he would speak faster and faster, tending to chop bits off words as he went, so you had to concentrate hard to follow his conversation. Many years later, I had occasion to interview him in a deliberately combative style for BBC radio, and he became very agitated indeed. I was concerned that the listeners would be left totally uncomprehending as technical words spilled out of him at an accelerating rate. An even more serious worry was that he might have a heart attack in the studio, as he was still recovering from triple-bypass heart surgery at the time. In the event, he lived for several more years.

Although Bohm was famed for his writing and philosophical works, particularly among readers of a mystical bent, he was a curiously isolated figure among the physics community. He was perhaps best known for his 1950s textbook on quantum mechanics. But very early on he decided he didn't like quantum mechanics as conventionally formulated, à la Bohr. So it became Bohm versus Bohr. Bohm thus took up the lonely torch of quantum dissent where Einstein had left it on his deathbed. With the help of a small band of devotees, most notably his Birkbeck colleague Basil Hiley, Bohm sought a theory in which the apparently random and unpredictable aspects of quantum phenomena had their origin in some deeper-level deterministic processes.

Bohm had this fascinating idea that, although some features of the world might look complicated, or even random, beneath it all there lies a hidden order, somehow "folded up." In later years, he was to call it "the implicate order." He used to perform an entertaining and instructive demonstration of folded-up order using a drop of dye deposited in a jar of glycerine. The jar had a handle that could be used to stir the dye into the glycerine, so that after a while it just looked like uniform gray gunk. But the apparent disorder of the dye is only illusory, because, if the handle is wound the other way, then—surprise, surprise—the dye "unmixes" from the glycerine and is restored to its original bloblike orderliness. In the smeared-out state, the order of the dye was merely hidden: it was "folded up." Might it be the case, I mused, that the random phases of electromagnetic waves—the thing that was puzzling me in my thesis work on the nature of time—represented a type of folded-up or implicate order?

Birkbeck College was located a few hundred yards from the Physics Department at University College, where I was studying, yet I only made the pilgrimage once. I carefully explained my project to Bohm, who listened politely. With some trepidation, lest I should be asking the great man a silly question, I ventured: "What is the origin of the random phase assumption?" To my astonishment and dismay, Bohm merely shrugged, and muttered: "Who knows?"

"But you can't make much progress in physics without making that assumption," I protested.

"In my opinion," replied Bohm, "progress in science is usually made by *dropping* assumptions!"

This seemed like a humiliating put-down at the time, but I have always remembered those words of David Bohm. History shows he is right. So often, major progress in science comes when the orthodox paradigm clashes with a new set of ideas or some new piece of experimental evidence that won't fit into the prevailing theories. Then somebody discards a cherished assumption, perhaps one that has almost been taken for granted and not explicitly stated, and suddenly all is transformed. A new, more successful paradigm is born. This happened when Einstein formulated the special theory of relativity. Everybody had assumed, without even thinking about it, that time is absolute and universal; the whole of classical physics was founded on this belief. But it was wrong—an unjustified assumption that brought Newton's laws of motion into conflict with electromagnetism and the behavior of light signals. When Einstein dropped the assumption, everything fell into place.

Anyway, I left my encounter with Bohm with the random phase assumption still disturbing me, so I decided to look up what Einstein had to say on the matter. In 1909, about the time he had been appointed associate professor at the University of Zurich, Einstein published a short note with Walther Ritz, a young but sickly physicist at the University of Göttingen in Germany.[1] Though sympathetic to the theory of relativity, Ritz thought Einstein did not fully understand the nature of electromagnetic radiation. He was convinced that there was an overlooked law of nature favoring retarded electromagnetic waves and suppressing the advanced variety. Ritz called this the "emission theory of light," because it distinguished between emission and its time reverse—absorption. He believed it was the explanation for the directionality in time that we observe in daily life.

Einstein disagreed. He insisted that the laws of electromagnetism must be symmetric with respect to time. The asymmetry of retarded waves came, he asserted, from statistical considerations. To see what Einstein meant by this, imagine a stone being thrown into a pond. It creates ripples that spread out from the point of impact and eventually die away in the shallows, lost among the waving reeds. These are retarded waves. A movie film of this sequence, run in reverse, would show advanced waves—ripples that appear around the edges of the pond and converge in an organized circular pattern onto a point. The latter scenario is not, strictly, impossible. It is conceivable, but highly unlikely, that the motion of the reeds would cooperatively and conspiratorially arrange themselves in such a way as to create just the right mix of little ripples to produce a precisely circular pattern of converging waves. The conspiracy involves many separate wave disturbances being choreographed to arrive in the middle of the pond at precisely the same time and in perfect step—i.e., with correlated phases. In reality, we would expect the chance motion of reeds to be largely uncorrelated, and the phases of the wavelets to be random.

Translating this into electromagnetic terms, we conclude that an advanced wave is not impossible, only extremely improbable. Imagine radio waves from a transmitter going off into the depths of space, where one day they may be absorbed by stardust or some other diffuse matter. The movie-in-reverse of this sequence of events consists of zillions of tiny radio emissions from across the cosmos coincidentally arriving at the same moment, in exact phase, at a point on Earth. Although the cosmic material certainly is a source of radio waves, and the Earth is bathed in this incoming electromagnetic noise, there is no evident correlation between waves emanating from, say, the direction of Leo and those coming from Pisces. To expect there to be so amounts to believing in a major cosmic conspiracy, whereby widely separated regions of the universe act cooperatively to send radiation towards us in carefully synchronized unison. But in practice the many different wavelets arrive with completely uncorrelated phases. Thus, according to Einstein, the retardation of electromagnetic waves stems from the random phase assumption!

The one-way-in-time nature of radio waves and other electromagnetic radiation forms but part of an elaborate matrix of physical phenomena that imprint on the universe an arrow of time. In daily life we have no trouble telling the direction of this arrow, because we are surrounded by processes that seem to be irreversible—human aging for one. But the ultimate origin of the arrow is a tantalizing mystery. In Chapter 1, I related out how Boltzmann believed he had found the source of the arrow in the second law of thermodynamics, only to have the basis of his proof kicked away by Poincaré. The relationship between the second law and the arrow of time was widely popularized by Arthur Eddington in the 1920s, and has been a beguiling topic of investigation ever since. A definitive answer to the mystery eludes us still.

The commandment that heat must always flow unidirectionally from hot to cold is clearly at the root of many "everyday" manifestations of the arrow. On a cosmic scale, this law describes a universe locked inexorably on the path of degeneration, sliding towards the final heat death. The fuel bill for the sun, incidentally, is about a trillion trillion dollars per second, 1993 prices. Of this, all but a few billionths streams off into the depths of space, wasted; the rest warms the planets. You can't get this energy back: it is lost irretrievably. Even as the solar system rhythmically chronicles intervals of time over the eons, cycles within cycles, a one-way clock lies ticking in the heart of the sun, totting up the fuel bill. No cycles here, just a mounting energy cost and dwindling reserves of fuel: irreversible, finite, inescapable. Eventually the sun will die, as will all stars, and perhaps the universe as a whole—unless it collapses to a big crunch first.

SIGNALS FROM THE FUTURE

In 1941, John Wheeler, the Princeton physicist, acquired a bright young student from New York named Richard Feynman, a man of colorful personality who was touched by genius and destined to become one of America's best-known and best-loved scientists. I confess I have always been amused by the thought of the collaboration between these two Americans, as different as chalk from cheese. Wheeler is a refined, patrician man, mild-mannered and impeccably polite. A colleague once said of Wheeler that he is a perfect gentleman inside of which lies a perfect gentleman. Feynman, by contrast, was famous for his brashness, irreverence, womanizing, practical-joking and bongo-playing.

Unlikely a match though they were, Wheeler and Feynman made a formidable team. They enjoyed a fruitful and cordial working relationship that extended over many years. In the midst of the Second World War, before both men got diverted to the Manhattan atomic-bomb project, they resolved to investigate the nature of time and the behavior of electromagnetic waves. Wheeler wanted to determine what would happen if advanced and retarded electromagnetic waves were always produced on an equal footing. This would mean, among other things, that a radio transmitter would send out half its wave power into the future, and half back into the past. It seems like a half-baked and pointless exercise to pursue.

In science, however, the mark of a really good idea is that what may appear crazy can come up trumps. You have to believe that either or both of them guessed the answer beforehand or they could have wasted a lot of time bogged down in calculational complexities. To most people, however, the result comes as a total surprise: it turns out that all the advanced waves disappear from view! This is why. When the retarded waves from a particular source on Earth, having spread out across the universe, encounter matter, they will be absorbed. The process of absorption involves the disturbance of electric charges by the electromagnetic waves, and as a result secondary radiation is produced by those faraway charges. This radiation too is one-half retarded and one-half advanced, in accordance with the assumption of the theory. The advanced component of this secondary radiation travels backwards in time, and some of it reaches the source on Earth. Naturally, this secondary wave is but a pale echo of the original wave, but a myriad of such pale echoes from across the universe can add up to a substantial effect. Wheeler and Feynman proved that under some circumstances the advanced secondary radiation can serve to double the strength of the retarded primary wave, bringing it up to full strength, while also canceling out the advanced wave of the original source by destructive interference.[2] At the end of the day, when all the waves and their echoes, backward and forward in time, are totted up, the net result is to mimic pure retarded radiation. Perhaps *this* is the reason for the arrow of time in the behavior of electromagnetic waves?

For Wheeler and Feynman's ingenious arrangement to work, it is essential that the matter in the universe is substantial enough to absorb all the radiation that flows out into space. In other words, the universe must be opaque to all electromagnetic waves. This is a stringent requirement. On the face of it, the universe appears almost completely transparent at many wavelengths. Indeed, we wouldn't be able to see distant galaxies if it were not. On the other hand, there is no time limit on the absorption process, because the advanced (backward-in-time) echoes can journey back in time and space from the very far future just as easily as from the near future. So the success of the theory turns on whether an outgoing electromagnetic wave would *eventually* be absorbed somewhere in the cosmos, maybe countless eons later.

We can't, of course, know if this will be the case, because we can't foretell the future, but we can extrapolate the present trends in the universe and make an educated guess. When this is done, the result seems to be in the negative—i.e., the universe is *not* completely opaque. That seems to finish off the Wheeler-Feynman idea, but there remains a more intriguing possibility. Suppose there is enough matter in the universe to absorb most of the radiation, but not all of it. According to Wheeler and Feynman, this would lead to an incomplete cancellation of the advanced waves. Might it be the case that there *are* some advanced waves "going into the past"—or, alternatively, coming from the future—but with such a low strength that we haven't spotted them yet?

In 1972, an American astrophysicist named Bruce Partridge went to the top of a hill to test this romantic conjecture.[3] He took with him a microwave transmitter with a large cone-shaped horn. On cloudless nights in August and September, he directed the horn skywards, taking care to avoid the Milky Way, and threw the transmit switch. The antenna beamed 9.7-gigahertz electromagnetic waves straight into space, in tight millisecond pulses. In the gaps between the pulses, the entire output was directed into a dead-end absorber attached to the apparatus. The system was thus designed to chop a thousand times a second between transmitting radio waves out into the universe—maybe to get absorbed in a trillion years—and into a screen and certain absorption an instant later. Partridge carefully monitored the power drain to see if it showed any hint of a millisecond variation. He then repeated the procedure placing a large absorbing screen in front of the horn, and checked that there was no difference in behavior.

The theory behind the experiment is that if any microwaves were getting beamed into the past, then, from a "forward-time" point of view, this would represent electromagnetic power flowing *into* the antenna rather than out of it. It would have the effect of supplying some energy *to* the apparatus to help pay for the energy conveyed up the horn and off into space by the normal "retarded" microwaves. If this were happening, there would be a slight difference in the power drain when the antenna was beaming into space from when it was beaming into the dead-end absorber. Unfortunately, Partridge found no trace of a millisecond wiggle in the power output, to one part in a billion. Clearly, backward-in-time radio transmission, if it exists, is exceedingly weak. Partridge estimated that only about 3 percent of the power would get absorbed by the atmosphere, and less than 1 percent by the galaxy: the rest would make it out into the vast intergalactic void. Whether these waves get absorbed eventually depends on the far future of the universe —even on its ultimate fate—about which we can only theorize. It could be that the universe *is* an efficient absorber of microwaves, and that Wheeler and Feynman's theory is right. Or it may mean that the theory is simply wrong, and that 100 percent of the waves emitted by microwave antennas are retarded. Whichever is the case, Partridge's experiment, and an improved version conducted some years later by Riley Newman, are the only examples of a cosmological experiment (as opposed to passive observation) in the history of science.

A rather different proposal to look for advanced effects was suggested in 1969 by an Oregon physicist, Paul Csonka.[4] His experiment concerns neutrinos rather than electromagnetic waves. The reasoning here is that physical objects acquire their sense of time direction by interacting with the world, so that things that interact only feebly may have only a weak sense of time orientation. In the case of neutrinos, their interaction with ordinary matter is so incredibly feeble that, according to Csonka, they may end up "losing their way in time" altogether. To give you some idea, a typical neutrino from the sun (a major local source) is affected so little by the ordinary matter of the universe that it is likely to travel a million trillion trillion light-years before being scattered or absorbed. The material universe is thus almost entirely invisible to it. Perhaps, then, it cannot "know" which way time is flowing in the great wide world, and so it may be inclined to "do things backwards" on occasions. Or so Csonka surmised. He suggested taking a careful look at beams of pions. These are subatomic particles that decay into, among other things, neutrinos. If Csonka is right, not only will a pion beam create a neutrino beam, it will also "attract" a sort of shadow neutrino beam from behind, consisting of "backward-in-time" neutrinos arriving in time for the pions to decay (the neutrino analog of Wheeler and Feynman's advanced radiation). These shadow neutrinos might, Csonka ventured, be detectable. Unfortunately, detecting any sort of neutrinos, shadow or otherwise, is a formidable challenge, as the feebleness of their interaction indicates, and, as far as I am aware, nobody has ever tried to look.

According to the writer Paul Nahin, Einstein himself took a passing interest in Wheeler and Feynman's theory following a seminar on the topic at Princeton University.[5] He pointed out that the basic idea had been around for years, and to prove it he dug out a paper by the German physicist Hugo Tetrode published in 1922. Which only goes to show that there is nothing new under the sun, even when it comes to time itself.

A MATTER OF TIME REVERSAL

Shortly after Wheeler and Feynman worked out their entertaining theory, Wheeler put Feynman on to another bizarre idea involving backward-in-time action. This time it had to do with antimatter. The concept of antimatter dates back to about 1930 and a famous prediction by Paul Dirac, who had been struggling to merge the new quantum mechanics with Einstein's special theory of relativity. Dirac wanted to know how a quantum particle such as an electron would behave when moving close to the speed of light. He discovered an equation that seemed to fit the bill, but he was baffled to find that every solution of the equation that described an electron came paired with a sort of mirror solution that did not seem to correspond to any known particle. After a certain amount of head-scratching, Dirac came up with a bold hypothesis. The "mirror" solutions, he claimed, correspond to particles that are identical to electrons, except their properties are reversed. For example, instead of having a negative charge, the mirror particles should be positively charged. Within a year or two, Dirac's "positrons" had been found in cosmic-ray showers. They really do exist.

Physicists eventually came to realize that every sort of subatomic particle in nature has a corresponding antiparticle. In addition to anti-electrons (still called positrons) there are antiprotons, antineutrons and so on. Today these antiparticles are routinely produced in the laboratory and are well understood, but in the 1940s they were still a bit mysterious. Only the positron was familiar. Positrons are created twinned with elec-trons in violent encounters between gamma rays and matter. Typically, a gamma-ray photon colliding with an atom produces an electron-posi-tron pair. The newborn electron flies off to enjoy a more or less perma-nent existence, but the poor positron faces hazards from the outset. If a positron runs into an electron (and the universe is packed with them), the pair will instantly annihilate, reversing the pair-creation process and giving back photons. This generally makes for a short career for the positron.

Now let me turn to Wheeler's proposal, as developed by Feynman. Fig. 9.1 is a spacetime diagram showing the creation and subsequent

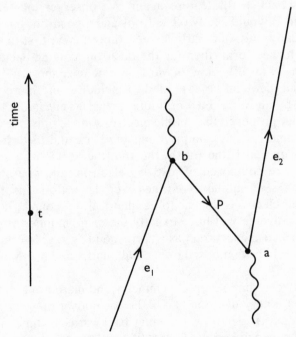

9.1 *The spacetime diagram depicts a photon creating an electron-positron pair* (e₂, p) *at a, with the positron subsequently being annihilated by electron e₁ at b. At time* t, *an observer would see three particles in existence: p, e₁ and e₂. According to Feynman, the zigzag track can be viewed as the world line of just one particle, an electron that travels back in time between b and a (see arrow).*

annihilation of a positron. The commonsense way to interpret this diagram is that the gamma-ray photon, depicted by the wavy line coming up from the bottom, creates an electron-positron pair at event A, the electron (labeled "e_2") goes off to the right, while the positron, labeled "p," heads left, hits a second electron (labeled "e_1") at event B and annihilates, creating a photon once more. The net effect is that electron e_1 has disappeared in one place to be replaced by electron e_2 in another. Feynman's audacious conjecture was that electrons e_1 and e_2 are really the *same* particle, even though in the interval between events A and B both electrons are present together!

Feynman's idea is that the continuous zigzag track in Fig. 9.1 should be seen not as a concatenation of three distinct particle world lines, but as a continuous spacetime path of a *single* electron. The backward-sloping segment of the track—the bit corresponding to the positron—then depicts the electron *going backwards in time*. This time-flip is denoted by the arrows on the world line. In its normal, electron phase, the arrow points forwards in time, but during the positron phase it points backwards. Viewed this way, the original undisturbed electron (e_1) emits a photon (at B) and bounces back in time, then absorbs a photon (at A) and rebounds back to the future again. An observer located in time between A and B would see two electrons and a positron, but Feynman says this is really only one particle seen three times: first (as e_1) in its original undisturbed form, then (as the positron) coming back from the future, and finally (as e_2) going forwards in time once more.

The essential idea can be extended to include many more electrons and positrons by allowing the world line to zigzag repeatedly (Fig. 9.2). In fact, Wheeler proposed that all the electrons in the universe are really one and the same particle, simply bouncing back and forth in time! In other words, you and I, the Earth, the sun, the Milky Way and all the other galaxies are composed of just *one* electron (and one proton and one neutron too) seen squillions of times over. This offers a neat explanation for why all electrons appear to be identical. Of course, it also predicts that the universe will have exactly the same number of positrons as electrons, because every zig has a corresponding zag. In other words, the universe would be composed of one-half matter and one-half antimatter.

The link between time-reversal symmetry and matter-antimatter symmetry is in fact a very deep one. Whether or not we take seriously the idea of positrons as electrons traveling backwards in time, it can be shown on quite general grounds that if the laws of the universe are strictly symmetric in time then the universe should have equal shares of matter and antimatter. Some cosmologists have suggested just that. Antimatter *looks* the same as matter, so you can't tell from casual inspection whether or not, say, the Andromeda galaxy is made of matter or

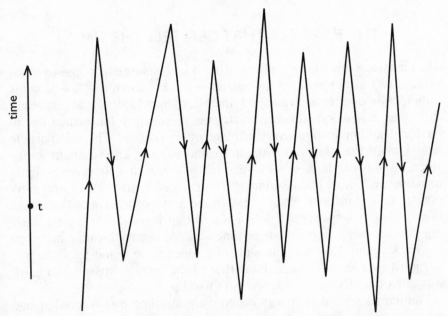

9.2 *Multiple zigzags could explain why all electrons are identical: they are all the same particle, bouncing repeatedly back and forth in time. An observer at time t would erroneously interpret the single world line as a multiplicity of disconnected segments.*

antimatter. Perhaps half the galaxies are in one form and half in the other? To test this fascinating possibility, astronomers have looked for ways that antimatter might betray its presence. Whenever antimatter encounters matter, prodigious quantities of gamma radiation are produced, with distinctive energies. There are many known examples of galaxies in collision with each other, so if half the galaxies are made of antimatter we might expect the universe to be replete with distinctive gamma rays. However, almost no gamma rays of the right energy have been found. This suggests a major predominance of electrons over positrons, and generally of matter over antimatter.

The conclusion we can draw from these observations—and it is a very profound one—is that nature is *not* symmetric between matter and antimatter, so the laws of the universe are *not* exactly symmetric in time. Whatever physical processes brought about the creation of the cosmic material, presumably in the extreme conditions of the big bang, they must have been lopsided in their relation to time, even if only slightly. In other words, there must be at least one basic physical process that is not exactly symmetric under time reversal.

THE PARTICLE THAT CAN TELL THE TIME

The idea that a fundamental physical law might violate exact time-reversal symmetry was not even on the agenda when Feynman was finessing his notion of antimatter and reversed time. But, coincidentally, at just about that time, a new subatomic particle was discovered that turned out to have crucial implications for the time-symmetry issue. The new particle was dubbed the "kaon." Although I had vaguely known about kaons since my days at high school—enough to know that they were highly unstable and short-lived subnuclear particles—I first sat up and took notice of them in 1966, when I read in the London press about a bizarre theory. The newspaper article suggested that kaons might occasionally flip into another universe, where time was running backwards, and then flip back again. This was the stuff of science fiction, and I was deeply intrigued, not least because the author of the theory, Russell Stannard, was one of my lecturers at University College.

Stannard's speculation was based on a startling discovery that had taken place two years previously, and which hinted that kaons could "do something funny" with regard to time. To explain just what requires a bit of background. When kaons were first discovered, in 1947, their existence was manifested by mysterious V-shaped tracks that formed in cloud chambers exposed to cosmic rays. From the start, physicists had suspected there was something odd about them. Kaons can be created by collisions between nuclear particles such as protons and neutrons, but once made they don't last very long. After a few nanoseconds, most will have decayed, mainly into pions. Both kaons and pions belong to a class of subnuclear particles collectively called "mesons." An important property that mesons share with protons and neutrons is that they interact with each other very strongly, meaning that reactions involving the transmutation of one type of particle into another tend to occur more or less instantaneously. This strong nuclear force is to be contrasted with another, entirely different, "weak" nuclear force. The weak force, extremely feeble by comparison, is responsible for many very slow nuclear processes, such as radioactive beta decay. To give an example, a typical strong-force interaction takes a trillion-trillionth of a second, whereas the decay of the neutron, which is induced by the weak force, takes about fifteen minutes.

All particles that are affected by the strong force are made up of combinations of smaller particles called "quarks." For example, protons and neutrons have three quarks apiece, mesons two quarks (strictly, a quark and an antiquark). There are probably six different varieties of quarks (five are definitely known), and the same number of antiquarks, so that there are thirty-six different combinations of a quark and an antiquark. This makes for a lot of possible mesons. The pion and kaon happen to have been among the first discovered, because they are the lightest. Kaons come in three varieties: electrically neutral, positively charged and negatively charged.

It was the way in which kaons decay that alerted physicists to their special peculiarities. A typical kaon is produced by the strong force, in a trice, following the collision of two strongly interacting nuclear particles. However, although the kaon decays into other strongly interacting particles (pions), it takes as long as a nanosecond to do so. This came as a shock. If a particle can be made in a trillion-trillionth of a second by a particular sort of process, why doesn't the particle decay in about the same time by the same sort of process? What goes forwards should go backwards. The situation is rather like throwing a ball in the air and finding it takes a million years to come down again. What is it that leads the kaon to take trillions of times longer to decay than it takes to be produced?

At stake here was an almost sacred principle of physics that had been accepted without question for as long as anyone could remember—the principle of the reversibility of all fundamental physical processes. A picturesque way of envisaging this principle is to imagine taking a movie of the process concerned and then playing it backwards. If the process is reversible, then the reversed movie should show a possible physical process too. Thus a film of a planet going around the sun, played backwards, would show a planet going around the other way. Nothing wrong with that. Of course, we have all spotted films being run backwards, because they show silly things like rivers flowing uphill and people walking backwards. However, these scenes involve complicated processes, and I am restricting myself for the moment to *basic* phenomena involving just a few elementary particles.

The reversibility of basic physical processes comes from the time symmetry of the laws that underlie them. This time-reversal symmetry is usually denoted by the letter "T." You can think of T as an (imaginary) operation that reverses the direction of time—i.e., interchanges past and future. Time-symmetric laws have the property that when the direction of time is inverted the equations that describe them remain unchanged: they are "invariant" under T. A good example is provided by Maxwell's equations of electromagnetism, which are certainly T-invariant. If you apply T to a retarded wave, you get an advanced wave—as I have already described. Advanced waves are physically possible, though for some reason we don't seem to see them.

You can invert time mathematically in a set of equations with no effort at all, but you can't so easily reverse the flow of time in the laboratory. However, T symmetry can be tested experimentally by reversing the *process* concerned—you can make all the things involved in the process go backwards, so this is really "motion reversal," which usually amounts to the same thing as time reversal. When you do that, you normally find the original process does indeed get reversed, and you end up back where you started, with the initial physical state restored. Furthermore, the reverse process goes at the same rate as the forward process.

Exact time reversibility was for decades taken for granted by physicists, for no very good reason. There was a vague feeling that something as simple as an elementary particle, or an electromagnetic wave, couldn't possibly have an intrinsic sense of past and future. The fact that the kaon seemed to flout this rule by taking trillions of times longer to decay than to be produced was therefore very strange—so strange, in fact, that the scientists involved in the discovery bestowed upon the kaon a new quality which they called "strangeness." Before long, other "strange particles" were found. The origin of their strangeness was eventually tracked down to a culprit: every strange particle contained a particular type of quark—a strange quark.

The reason for the strange behavior of the strange particles soon became clear. Here is the gist of it as we understand it today. A strange particle is produced when a strange quark is created (among others) in a high-energy collision of nuclear particles. The collision also produces a strange antiquark. Because the antiquark has "antistrangeness," there is no net production of strangeness involved in this process, so it could go in reverse, with the quark-antiquark pair annihilating once again. However, the quark-antiquark pair immediately get separated, with the strange quark locked up in the kaon. The kaon is therefore unable to decay unless by chance it meets a stray particle containing an antistrange antiquark—most unlikely in practice. The process would be permanently irreversible except that the weak nuclear force is capable of changing one type of quark into another. In particular, it can change a strange quark into one of the common-or-garden variety of nonstrange quarks. Once this has happened, a decay route lies open for the kaon. But the weak process is very tardy, which is why the kaon takes (relatively) so long to decay. The upshot is that the production and decay processes of the kaon are not really the inverses of one another after all, and the law of time reversibility is not violated by these strange processes.

As it turned out, this was just the beginning of the story. There was still something weird about the electrically neutral kaon, denoted K^0. When physicists tried to measure how long they take to decay into pions, they were astonished to discover that these kaons seem to have two quite different lifetimes. Sometimes they decay into two pions after about ten-trillionths of a second, other times they decay into three pions with a lifetime thousands of times longer than this. It was almost as though there were two different identities inhabiting the same particle—a sort of Jekyll and Hyde existence.

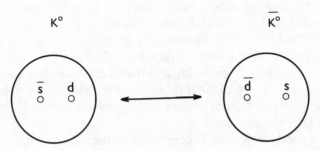

9.3 Identity crisis. The neutral kaon, K^o, is made of two smaller particles: an antistrange antiquark ($\bar{s}$) and a down quark (d). The antiparticle, $\bar{K}^o$, consists of a strange quark (s) and an antidown antiquark ($\bar{d}$). The weak force can change d into s and $\bar{s}$ into $\bar{d}$, and vice versa. This causes the kaon to vacillate in identity between K^o and $\bar{K}^o$.

An explanation for this further puzzle was soon forthcoming. The K^o consists of an antistrange antiquark bound to a quark of another type called "down" (see Fig. 9.3). The weak force can change the down quark into a strange quark, and at the same time change the antistrange antiquark into an antidown antiquark. The net effect of these two transmutations is to turn the K^o into its antiparticle, denoted $\bar{K}^o$. This spontaneous change in identity can go the other way too, taking the $\bar{K}^o$ back to a K^o. The neutral kaon thus has a sort of permanent identity crisis: it doesn't know whether it is a K^o or its antiparticle $\bar{K}^o$, and it keeps oscillating back and forth between the two. This rapid mixing of identities means that physicists monitoring kaon decay were actually being presented with a hybrid entity: a kaon-antikaon. In fact it is more complicated than this. There are *two* such entities, because the mixing of K^o and $\bar{K}^o$ can occur in two distinct ways, according to whether the mixed system is symmetric or not under *space* reflection. (By this I mean that, if the two mixtures were viewed in a mirror, one would be the inverted image of the original and the other would not.) Having discovered the existence of these two distinct mixing modes, physicists could understand why there were two different decay schemes. Mixture 1, better known as K_1, is unchanged in a mirror, so it had better decay into an *even* number of pions, this also being symmetric as far as mirrors are concerned. On the other hand, mixture 2, called K_2, reverses in a mirror, so it should decay into an *odd* number of pions. There are thus two completely different decay schemes, one into two pions and another into three, depending on which mix of K^o and $\bar{K}^o$ happens to exist when the decay occurs. Because the three-pion process is slower, that decay route has a correspondingly longer lifetime.

To understand fully the significance of the two decay schemes, you have to take into account that there is a rather basic link between mirror reflection and time reflection. For example, a spinning sphere viewed in a mirror spins backwards, and looks exactly as if its motion had been reversed instead. It can be shown quite generally that mirror reflection is physically equivalent to time reversal—with one refinement: you have to swap the identities of particles and their corresponding antiparticles. So the existence of two distinct two-pion/three-pion decay schemes, by preserving mirror-reflection symmetry, neatly expresses the invariance of nature under time reversal.

So far so good. However, it came as a further shock when, in 1964, a group of Princeton physicists led by Val Fitch and James Cronin found that one in every few hundred K_2 particles decayed into *two* pions instead of three! I was still at high school at the time, and I can remember learning of the discovery through the unlikely medium of the guest of honor's address at the annual school speech day. The implications of the Fitch-Cronin experiment were nothing short of iconoclastic (which was why the said speaker deemed it a fitting comment for a formal occasion). It was soon established that the maverick behavior of the kaon implied the violation of the hitherto sacred principle of time-reversal symmetry.

A possible way to think about how the kaon violates T symmetry is this. The K_1 and K_2 states arise, as I have explained, as a sort of hybrid or mixture of kaon and antikaon. Envisage the particle rapidly flipping back and forth in identity: kaon-antikaon-kaon-antikaon . . . One can ask whether these flips are *perfectly* symmetric—i.e., whether the rate of going from kaon to antikaon is exactly the same as the rate of going from antikaon to kaon. If not, the hybrid entity might linger longer as a kaon than an antikaon, or vice versa. Everyone assumed that, as the laws that induce the kaon-antikaon flips should be exactly symmetric in time, nature ought not to distinguish one process from its inverse, and the two rates should match precisely. But there is a tendency for the kaon to spend more time as a $\bar{K}^O$ than as a K^O.

This unexpected behavior implies that the kaon possesses an *intrinsic* sense of "past-future." Although the effect is tiny, it is deeply significant, and deeply mysterious—hence the wild speculation of Russell Stannard to explain it in terms of the kaon popping off to visit temporarily a time-reversed parallel universe. *Scientific American* columnist Martin Gardner commented: "Stannard's vision bifurcates the cosmos into side-by-side regions, each unrolling its magic carpet simultaneously (whatever 'simultaneously' can mean!) but in opposite directions."[6]

You haven't heard from me for quite a while, but now I'm hopelessly confused about something. I thought that Einstein had finished off the concepts of past and future. How can physicists claim that kaons have an inbuilt sense of past-future asymmetry?

Good point. We have a problem with language here. Einstein ruled out the absolute division of time into *the* past and *the* future separated by a universal present moment or "now." But this does not preclude our distinguishing in an absolute way between past and future *directions* in time. We use the words "past" and "future" in two subtly different ways here. A similar difference crops up in our use of the words "north" and "south." We frequently talk about "the North" or "the South" when referring to *places*, as well as "north" and "south" meaning *directions* in space. In America, "the South" means states like Alabama and Texas; in Britain, "the North" is associated with cities such as Manchester and Newcastle. There is even an asymmetry between north and south, occasioned by the fact that the Earth is rotating. This asymmetry is denoted by the arrow on the compass needle, which plays a role for spatial asymmetry analogous to the arrow of time. The humble kaon is able to tell the time in a limited sense: it knows the difference between the two *directions* of time, past and future. But in no way does the kaon *divide* time into past, present and future.

THE LOPSIDED UNIVERSE

The minute lopsidedness in time that infests the subnuclear realm carries with it a related lopsidedness in connection with matter and antimatter. Recall that the T violation can be traced to the fact that the rate at which kaons turn into antikaons does not precisely balance the reverse process, antikaons into kaons. If such an asymmetry exists between matter and antimatter, even at the minuscule level observed, it could offer a natural explanation for why the universe is predominantly made of matter. We can imagine that most of the material in the universe was produced in the hot big bang. Initially there was an explosive mix of matter and antimatter, but the proportions were not quite even: there was a slight excess of matter due to the T-violating effects. The mixture would not have survived for more than a second or so before wholesale annihilation converted almost everything into gamma rays. This would have removed all the antimatter, but the tiny excess of matter would be left over, unscathed. It was this excess that went on to make the galaxies, whereas the gamma rays, greatly weakened by the expansion of the universe, became the cosmic-background heat radiation. If this theory is correct, then our own existence can be seen to hinge crucially on the minute temporal wonkiness that nature allows. It is an asymmetry so small as to be almost an afterthought, yet without it we would not be here.

Once the idea of T violation had sunk into the minds of dumbfounded physicists, the search began in earnest for the most sensitive way to

measure it. The place where this search has been pursued most assidu-
ously lies in a spectacular river valley in southeastern France, not far
from the fashionable Alpine ski resorts. Here is found the old town of
Grenoble, birthplace of the famous musician Hector Berlioz, he who
once remarked wittily: "Time is a great teacher, but unfortunately it kills
all its pupils."[7] It is also the site of a major nuclear physics laboratory.
The French scientists have directed their attention, not at kaons, but at
the humble neutron, which may have concealed in its electromagnetic
properties a vital clue to the temporal lopsidedness of nature.

You would be forgiven for supposing that neutrons, being electrically
neutral, had *no* associated electromagnetic properties. Most physicists
assumed as much when the neutron was first discovered. It therefore
raised a few eyebrows when, in 1933, the German physicist Otto Stern
discovered that a neutron acts as if it contains a tiny bar magnet. Today,
the origin of this magnetism doesn't seem so mysterious. We know that,
although the neutron is electrically neutral overall, it is not a point particle but a composite body containing three electrically charged quarks.
The total charge sums to zero, but the quarks can create a magnetic
field, because all neutrons are found to be spinning. You can think of
each neutron as a little ball rotating on its axis like a tiny planet, except
that every neutron spins at exactly the same rate—the spin is a fixed
quantity for the neutron, like its mass. Closer inspection reveals a more
complicated picture: it is the charged quarks within the neutron that are
spinning, each constituting a minute electric current which creates a
magnetic field. The overall effect is to produce a net magnetic field
aligned along the neutron's spin axis with the form known as a "dipole."
This name derives from the fact that, as in a bar magnet, there is a north
pole at one end and a south pole at the other.

The existence of charged particles inside the neutron opens up another
possibility. The spin axis of the neutron defines a fixed direction in space.
Even though the total charge of the neutron might add up to zero, it may
be that the positive charge prefers, on average, to congregate in one
region relative to the spin direction, the negative charge in another. This
would create an *electric* dipole field. If the neutron can have a magnetic
dipole, might it not have an electric dipole too?

This is where the arrow of time comes in. Imagine playing a movie
film of the neutron backwards. Nothing much changes, except that the
neutron now spins in the opposite direction. On the other hand, if the
neutron has an electric dipole, it would not be affected by time reversal,
because it doesn't depend on the quarks' motions, only their positions.
Therefore, reversing time has the effect of flipping the relative directions
of the spin and the electric dipole. This is most readily seen by using a
diagram (Fig. 9.4). The direction of spin can be labeled by analogy with
the Earth's rotation. As drawn, the neutron has its "northern hemi-

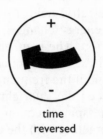

time
forward

time
reversed

9.4 Glimpsing time-symmetry violation. The neutron spin flips direction when time is reversed. The electric dipole is unaffected.

sphere" at the top, where the + charge is located, but in the time-reversed version the northern hemisphere lies at the bottom, with the − charge.

The opposing relationship of spin and electric dipole under time reversal would show up if the neutron was immersed in an external electric field. The electric field would act on the electric dipole and try to twist the neutron so that the + end lies towards the − of the field and vice versa. This interaction involves a certain amount of energy. If we could look at a particular neutron and reverse the direction of time, the neutron would spin the other way, but the dipole and external electric field would remain unchanged. Though we cannot actually reverse time, we *can* reverse the external electric field. This leaves the spin direction unchanged, but would change the electric-interaction energy with the dipole (because the + and − would effectively flip relative to the external field). This is completely equivalent to time reversal, because all that matters is the *relative* direction of the spin and electric dipole. So you can test for time reversal by reversing the applied electric field and seeing if the energy of the neutron changes.

It is worth pointing out that the same cannot be done for the magnetic field of the neutron. As explained, the neutron's magnetic dipole arises from tiny electric currents inside the neutron, and if time were reversed the direction of these currents would also reverse. Unlike the case of a static electric dipole, a magnetic dipole caused by spinning charges flips under time reversal along with the spin direction. Thus the relative orientation of a magnetic dipole and the spin arrow remains the same when time is reversed. Any interaction energy with an external magnetic field would also be unchanged. The conclusion of this chain of reasoning is that the existence of an electric-dipole moment in the neutron would be a sign that the world is *not* symmetric under time reversal. In other words, if the neutron possessed an electric dipole, however small, it would have an inbuilt sense of the direction of time.

To measure an electric dipole in practice, you put your neutron in a powerful electric field, reverse the direction of the field, and see if the neutron's energy changes. To monitor the energy, you provide a magnetic field too. The neutron tries to twist and align in the magnetic field. To assist it, a radiofrequency electromagnetic wave is directed at the neutron, and if the frequency of the wave just matches the energy difference between the "upright" and "inverted" spin states, then the wave induces the spin direction to flip. Any extra energy due to the electric dipole interacting with the electric field should show up in the fine-tuning of the radio wave. This experiment is a very sensitive test of time-reversal symmetry.

So far, no such violation has been discovered. According to the French experts, if the charged quarks distribute themselves in a lopsided way inside the neutron, then the average separation of the region of positive charge from the region of negative charge must be less than 10^{-25} centimeters, or a ten-trillionth of the size of the neutron. This is already awfully small, though the scientists are not discouraged. Many popular theories of particle physics involve T violation, but they suggest that time symmetry should be broken at a level just a little bit beyond the power of the French experiment to detect. At a much weaker level still, the same weak force that induces kaons to decay should also affect neutrons, so a sufficiently sensitive measurement must detect an electric dipole.

The expectation that time-reversal symmetry has to be broken at some level has spurred on experimenters all over the world to seek out tiny electric dipoles, not just in neutrons, but in atoms and molecules too. Current favorites are mercury and thallium fluoride. Experiments with molecules promise to be much more sensitive than those with nuclei, and should enable physicists to tease out evidence for T violation in due course. A group at Yale University hopes to be able to detect dipoles of as little as 10^{-28} centimeters across using the exotic molecule ytterbium fluoride (YbF).

The significance of a positive result in these dipole experiments would be that a fundamental particle such as a neutron—a constituent of ordinary matter—would have an *intrinsic* time orientation. By extension, the material contents of the universe would possess a tiny but nevertheless important sense of the direction of time. Past and future would be etched into the structure of matter at a basic level.

That's pretty astounding! Past and future in the universe are related to beginning and end. How can a tiny particle like a neutron or a kaon know about the big bang and the cosmic origin? There isn't any signpost in time saying "This way lies the big bang," is there?

Actually, there is. The expansion of the universe defines a temporal direction away from the big bang towards the future.

You mean kaons are tuned in to the cosmos? They can sense the expansion of the universe? That seems frightfully clever for a humble subatomic particle.

It does indeed. But no less a physicist than Yuval Ne'eman, who helped lay the foundations of the quark theory of matter, suggested as much in 1970. He claimed that the direction in time attached to kaon decay is directly linked to the cosmological motion. Therefore, if the universe was contracting rather than expanding, the time asymmetry of kaon decay would be the other way around. In effect, "a contracting matter universe is the same as an expanding antimatter universe."[8]

But how can a kaon, or any other subatomic particle, know what the universe is doing?

It all goes back to gravity. It was Einstein's theory of gravitation that gave us the possibility of an expanding universe. Perhaps there is some ill-understood aspect of gravitation that relates to T violation? After all, gravitation gives us one of the most conspicuous arrows of time— namely, black holes. You can fall into a black hole, but you can't get out again. Similarly, the formation of a black hole from the collapse of a star is an irreversible process. Unfortunately, there have been some sharp disagreements by the experts on this topic. In 1974, Stephen Hawking was propelled to fame by his discovery that black holes aren't black, but glow with quantum radiation. Small black holes get very hot indeed, and eventually evaporate away in an explosive release of energy. Careful mathematical analysis showed that the black hole acts like the ultimate randomizer: if ordered matter falls in, its energy comes back out again completely laundered, in the form of totally disordered radiation with perfectly scrambled phases. (Those random phases again!)

The Hawking effect advertised a unique arrow of time: order to disorder, care of the black hole. But Hawking himself viewed things differently. Shortly after the term "black hole" became fashionable, people began talking about white holes. What are they? Well, black holes in reverse. Instead of swallowing things up voraciously, they spew them out. White holes are not known to exist, and most scientists dismiss them out of hand, like all time-reversed contrivances. The way Hawking saw it, however, is that, if black holes emit radiation, they look rather like white holes. Confined to a box in thermodynamic equilibrium at a constant temperature, a black hole and a white hole would seem to be indistinguishable. One person who objected to this intriguing claim was

Roger Penrose, who insisted that a black hole and a white hole are completely different. In the mid-1970s, he and Hawking would argue out their respective cases in entertaining public encounters.

Penrose believes that gravity holds the key to the arrow of time: that there is an intrinsic lopsidedness to time when it comes to gravitational fields. At least there is when those fields are situated in the vicinity of spacetime singularities of the sort that exist at the centers of black holes (and white holes) and the big bang (and big crunch). Penrose concedes he doesn't know the origin of this lopsidedness, but thinks it might somehow link up with the kaons' T violation.

Does that mean, if the universe starts to contract, the arrow of time will reverse?

Ah, there's an interesting question! Now read on . . .

CHAPTER 10
BACKWARDS IN TIME

Time will run back, and fetch the age of gold.

JOHN MILTON

A reversal of the arrow would render the external world nonsensical.

ARTHUR EDDINGTON

INTO REVERSE

But I unfortunately was born at the wrong end of time, and I have to live backwards from in front.

T. H. WHITE

The idea that time can run backwards may seem amazing, but it is far from new. When you stop to think about it, any belief that time is cyclic must involve "going into reverse" at some stage, so that the world can be returned to its initial state. Plato gave a graphic description of this phase, seen through the eyes of an imaginary stranger:

> The life of all animals first came to a standstill, and the mortal nature ceased to be or look older, and was then reversed and grew young and delicate; the white locks of the aged darkened again, and the cheeks of the bearded man became smooth, and recovered their former bloom; the bod-

ies of youths in their prime grew softer and smaller, continually by day and
night returning and becoming assimilated to the nature of a newly born
child in mind as well as body; in the succeeding stage they wasted away
and disappeared.[1]

In the 1960s, the astrophysicist Thomas Gold came up with a theory
rather like this. He arrived at it by reflecting on the uncontentious fact
that the really important arrow of time in the universe is the flow of heat
away from the sun and stars into space. This, reasoned Gold, is the basic
process that imprints a past-future asymmetry on the world. Here lies
the arrow of time!

Gold sought to identify the ultimate origin of the arrow by asking *why*
there is a one-way flow of heat from stars into space. What causes it? A
naïve answer springs immediately to mind: stars are hot, but outer space
is cold. Heat flows from hot to cold, in accordance with the second law
of thermodynamics. But one can go on to ask *why* the universe is cold
and dark. The answer to that one has something to do with its expansion.
The bigger the universe gets, the more heat it can soak up. "It is like
pouring water into a barrel which fails to fill up, not because it has a leak
but because it is increasing its size all the time," explained Gold.[2]

To make his point, Gold invited us to imagine a perfectly reflecting
box magically placed around the sun, isolating it from the wider universe. The contents of the box would eventually reach thermodynamic
equilibrium, settling down at some very high uniform temperature. No
more heat would flow away and be wasted; all the energy would be
trapped and retained. The sun would then remain like this forever, with
the arrow of time snuffed out. Now, if someone were to drill a little hole
in the box to let some radiation escape, the balance of thermodynamic
equilibrium would be upset; heat would start to flow again, and irreversible change would once more occur; the arrow of time would be temporarily restored. Close off the hole, and the arrow will obligingly fade
away. So the arrow of time associated with the flow of energy out of the
sun depends on its being able to dump its heat unrestrictedly into the
cold vastness of space.

If the universe were static and the stars had been shining steadily for
long enough, the universe as a whole would fill up with heat and light
radiation in much the same way as Gold's box, but on a grander scale.
Radiation would accumulate in the space between the stars, slowly getting hotter and hotter, until eventually the cosmos would be fiercely hot
everywhere; there would be no dark, cold spaces left. From Earth, the
whole sky would appear to glow like a furnace. Eventually equilibrium
would prevail and the universe would reach a uniform high temperature:
a heat death. No further change would occur, for better or worse. This
state of affairs has not come to pass because the universe is not static; it

is expanding. As fast as the stars try to heat the universe up, space expands some more and keeps it cool. Moreover, the stars have not been shining for long enough to dump truly large quantities of heat into space: the universe originated only a few billion years ago.

Once he had established the connection between the arrow of time and the expansion of the universe, it was but a small step for Gold to suppose that, if the universe were to start contracting at some stage, the arrow would be reversed. Then "radiation would tend to converge onto objects and heat them; heat in general would flow from cold bodies to hot bodies," he suggested.[3] In other words, time would "run backwards."

Gold had in mind a cosmic cycle lasting tens of billions of years (see Fig. 10.1). Reversal would not take place for eons yet, by which time Plato's bearded men and youths would doubtless be a long-vanished memory. Nevertheless, the prospect that the universe might neither die nor go on degenerating forever, but might somehow rewind itself, is deeply intriguing, even if there isn't anybody around to watch.

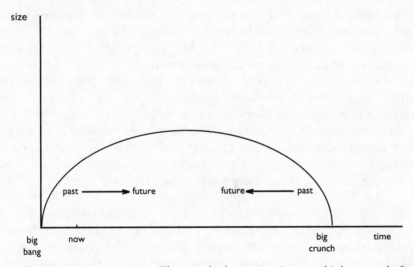

10.1 Time-reversing universe. The graph shows a universe which expands from a big bang, reaches a maximum size, and collapses to a big crunch. Time runs forwards in the first half and backwards in the second; its direction is denoted by the arrows. Because of the symmetry, the words "bang" and "crunch," and "past" and "future" are interchangeable. Our own epoch is indicated, using the time convention adopted.

THINKING BACKWARDS

It's a poor sort of memory that only works backwards!

LEWIS CARROLL

Of course, if there *was* someone about, it is fascinating to consider what he or she would actually experience. The arrow of time is so powerful and pervasive that its reversal would leave any being stuck with forward-time perception nonplussed and helpless. Imagine witnessing broken eggs reassembling themselves as if by a miracle, water running uphill, snow unmelting into snowmen, water in unheated pans spontaneously boiling, and so on. These processes would not merely seem unnerving and surprising, they would strike at the very heart of rationality. Prediction and memory play a vital part in all our activities, and a being who found these faculties operating the wrong way relative to the outside world would be utterly helpless.

The asymmetry imposed on the world by the laws of thermodynamics implies a *logical* directionality too. For example, I know that, if I leave a hot drink for an hour or so, it will be cold when I inspect it. But I cannot be sure that a cold drink was hot one hour previously. It might have been hot ten hours before, or concocted cold in the first place. Both hot and cold drinks become cold drinks one hour later. So the hot-to-cold reasoning doesn't run backwards. Many initial states lead to the same final state; retrodiction has no unique answer. The logical arrow is similar to arithmetic. Everyone can deduce that $12 + 15 = 27$, but we are stymied by the question: "What makes 27?" Going backwards uniquely from the answer to the question is in general impossible: the number 27 might be reached from $10 + 17$, or $3 \times 3 \times 3$, or some other way.

The concept of causality also has a strong directional flavor. We like to suppose that cause precedes effect. We would be uncomfortable with the idea that a breaking window causes a stone to be thrown, or a shadow passing across the Earth causes the moon to move in the way of the sun. It would be hard to make sense of a world where cause and effect were interchanged. Even with practice, prediction would be a hazardous affair in a time-reversed world. Imagine sipping a cold drink unsure of whether the liquid would stay cold, or start to boil spontaneously in your stomach. In fact, as the writer and mathematician Martin Gardner has pointed out, life in a time-reversed world would be even trickier than that.[4] An individual whose bodily and sensory functions are entirely reversed would be essentially stranded. For example, he would be unable to see or hear anything, because all light and sound waves would be leaving his organs and traveling back to the objects that emitted them.

This nightmarish scenario is, however, unlikely to occur. Our brain processes depend on the same physics as the rest of the universe, so they too would be reversed in a time-reversed world, along with the stream of consciousness and the memory and reasoning processes that attach to them. In other words, we would perceive and think in reverse in such a world. Our mental activity, including logical reasoning and concepts like causality and rationality, would all be inverted too. So a time-reversed being would not feel time-reversed at all. To it, all would appear normal.

It might seem as if I have defined time reversal away into a meaningless relabeling of past and future, but that is not so. It still makes good physical sense to talk about one region of the universe having its time direction reversed relative to another, even though the respective inhabitants of these regions would each regard their portion of the universe as "normal," and the other as "reversed." It is, then, instructive to contemplate what these respective groups of beings might have to say to each other about their relative experiences. Norbert Wiener, the inventor of the science of cybernetics, considered the communication problems that would ensue should this state of affairs exist. Imagine an attempt at a conversation between us and an alien being—perhaps in a neighboring star system—whose time sense ran the other way from ours. Wiener observed that the directionality of the alien's logic would be flipped, transforming his rational message into gobbledygook:

> Any signal he might send us would reach us with a logical stream of consequents from his point of view, antecedents from ours. These antecedents would already be in our experience, and would have served to us as the natural explanation of his signal, without presupposing an intelligent being to have sent it.[5]

In other words, normal concepts of meaning and explanation would be turned on their heads, precluding any intelligible exchange of information. Notions like chance and order would not translate. The alien's information would become our entropy, and vice versa. Thus: "If he drew us a square . . . [it] would appear to us as a catastrophe—sudden indeed, but explainable by natural laws—by which that square would cease to exist." Wiener concludes that, within any world in which communication is possible, the direction of time must be the same throughout. This conclusion is very frustrating. After all, the time-reversed alien would know our future and we would know his. He would remember all the disasters that lie in our future, but be powerless to warn us!

ANTIWORLDS

They spend their time mostly looking forward to the past.

JOHN OSBORNE

How seriously can we take the idea of different regions of spacetime having opposed arrows of time? Surprisingly, it is a recurring theme in physics and cosmology. One context where it keeps popping up is in connection with black holes. In Chapter 4, I mentioned how Finkelstein, Kruskal and Szekeres discovered in the late 1950s that the surface of a Schwarzschild black hole is not a physical barrier, but merely a gateway to a weird spacetime region beyond. Just how weird became clear when the algebra describing this region was scrutinized.

One can use the mathematics to calculate what a hapless astronaut—Betty, for instance—would see *inside* the black hole if she fell into one. We know she would soon come to grief at the central singularity, but on the way to her appointment with destiny she would be able to take note of her surroundings. This would include part of the region inside the black hole, and also the outside universe whence she originated. As I explained in Chapter 4, although Betty can't be seen from outside the hole, the converse isn't true: some light from outside the hole would follow her in and overtake her before she hit the singularity, so she could still see the outside world that she had so recently left. But there is more. Once inside, Betty would be able to glimpse another, completely different, region of spacetime lying in some sense "on the other side" of the black hole, a region that is totally inaccessible to us.

According to the idealized mathematical description, the "other universe" is a mirror image of our own, stretching away to infinity. There is, however, a crucial difference. The direction of time in the other universe is reversed relative to ours. This would make for a bizarre set of experiences as Betty plunges inwards, because she would be witnessing two different universes with opposite arrows of time. In fact, the region inside the black hole would be a melting pot of oppositely directed influences, a confused region where time-forward and time-reversed intersect and clash. However, although Betty can see the other universe, she can't travel into it any more than she can return to our universe. She is trapped by the intense gravity and drawn irresistibly on to the singularity. Clearly, the Schwarzschild black hole does not offer a way to put Plato's scenario to the test. Nevertheless, the idea that there may exist a sort of parallel universe with time running the other way—an antiworld, if you like—has a certain appeal. We have encountered such a speculation before, in the context of the kaon.

When pressed, most physicists and astronomers would dismiss the Schwarzschild antiworld as a fiction, and for good reason. Unless the universe originated with black holes already threading it, then no antiworld would exist. This is because the solution of Einstein's equations on which it is based applies only to the *empty* region outside matter. If a black hole forms from the collapse of a star—the usual scenario—then this solution cannot be continued through as far as the antiworld, because the matter blocks the way.

There are more general problems about patching together regions of the universe with oppositely directed arrows of time. For example, what happens at the join? To appreciate the sort of mayhem that ensues, imagine a simple thing like a game of snooker. Suppose that a mad scientist creates a laboratory in which time is made to run backwards, and equips it with a snooker table. In normal play, the propelled cue ball strikes an ordered triangle of target balls and scatters them into a disorderly state. Run in reverse, chaotically moving target balls somehow contrive to arrive in a triangle simultaneously, and trade impacts in such a way as to come to rest, concentrating their divested energy into the cue ball, which then races off down the table. This might be the scene that the scientist would observe by peeking through the lab window during a fragment of the game. The coming together of the balls in this strange way is exceedingly vulnerable to the slightest disturbance. A merest perturbation to just a single ball's motion would compromise the exquisite choreography and destroy any hope of orderly assembly into a triangle. (If you are not convinced, go ahead and try doing the experiment.)

The hypersensitivity of a reversed-time system implies that random influences intruding from the outside universe would soon wreck the experiment. If the lab could be *completely* sealed off, then time reversal would in principle be possible. But in practice this cannot be done. Heat and gravitational disturbances will always get through to a limited extent, nudging the contents of the lab with minute but fateful tugs, and destroying the delicate orchestration. Molecules are far more sensitive to disturbances than snooker balls. Even the odd photon coming into the imaginary lab through the inspection window may be enough to bring about dramatic change. Once a single disturbing influence gets in, the knock-on effects cascade uncontrollably, rapidly amplifying the original disturbance, until the influence encompasses everything in the lab, snooker balls included.

Chaos at the interface scuppers the entertaining theory, mooted in Chapter 9, that there may exist star systems where time runs in the reverse sense of our own. Recall that we would not see such objects, because light would be traveling *from* our eyes to those stars, raising the possibility that they are lurking out there invisibly in space. Alas, the muddling entanglement of their advanced starlight with our retarded light would shatter the unstable arrangement, and force the dominance of one arrow of time over the other (which one emerged victorious would depend on circumstances). This is also the conclusion reached by theorists studying white holes. Suppose a white hole formed in the big bang, surrounded by a region of the universe where time had its normal direction. Incoming photons and other disturbances would rapidly produce an instability, and convert the white hole to a black hole in short order.

WINDING THE CLOCK BACK

You can never plan the future by the past.

EDMUND BURKE

None of the foregoing problems means it is strictly impossible for neighboring regions of the universe to have mutually reversed arrows of time. The crazy-lab scenario described above involves the enforced juxtaposition of two such regions, which are overwhelmingly likely to be incompatible and lead to chaos. However, as I mentioned in Chapter 1, Boltzmann was already speculating about time reversal over a century ago, in a way which circumvents the above-mentioned difficulties.

Boltzmann appreciated the key role of chance in molecular activity. Among a collection of chaotically moving particles, there is always a small probability that some of them will be found to be blindly cooperating, perhaps coming together like the snooker balls to form an orderly arrangement. Statistics reveal that the odds against such accidental "conspiracies" soar as more and more particles are included. For example, from time to time a collection of ten oxygen molecules careering about in a flask will sometimes be found, by chance, to be located in the right half of the flask, leaving the left half empty. Typically, this will occur about once a second. But you could expect to wait many minutes for twenty molecules to do it. When you take into account that a 100 liter flask of air contains more than a trillion trillion molecules, it is no surprise that we don't see such improbable events on an everyday scale. Nevertheless, given long enough, it could happen. The work initiated by Boltzmann, and elaborated by Willard Gibbs, Einstein and others, confirmed that small-scale reversals are always occurring at the molecular level, over very short durations of time. In his 1905 paper on Brownian motion, written in the same year as his paper on relativity, Einstein investigated how a tiny particle suspended in a fluid can get knocked about as a result of the uneven molecular bombardment of its surfaces. In effect, molecules on one side can "gang up" and kick the particle harder than those on the far side, causing it to jerk about. These tiny motions reveal hidden fluctuations going on all the time in the fluid, amounting to the merest hint of a Boltzmann-style entropy reversal. Noticeable, human-scale reversals are staggeringly improbable. But if the universe were truly infinitely old, and otherwise unchanging in the large (Boltzmann himself knew nothing of the expansion of the universe), eventually substantial reversals in time must occur.

In Chapter 1, I alluded to Boltzmann's astounding suggestion that the universe achieved its present order as a result of a stupendously rare cosmic-scale fluctuation. The idea here is that for near-eternity the universe languishes in a dismal state very close to thermodynamic equilibrium—the famous heat-death condition—in which there is no arrow of time and nothing much of interest happens. However, very occasionally it will, purely by chance, stir itself and produce spontaneous order. After an immensity of time, during which countless random fluctuations come and go on small and intermediate scales, there must eventually occur a fluctuation of truly astronomical size, a concordance of mind-numbing proportions, involving countless trillions of particles blindly arranging themselves into stars, planets, people, etc. During this "winding-up" phase, there will be an arrow of time that points backwards. When the fluctuation is complete, the universe proceeds to unwind again, gradually slipping back to its more normal equilibrium state, generating a forward-pointing arrow as it goes. This incredible model describes a type of pseudo-cyclicity, because such spontaneous ordering and disordering episodes will occur infinitely often in the infinite duration of an everlasting universe.

A key feature of Boltzmann's time reversals is their back-to-back quality. The arrows point *away* from each other—i.e., ours points to the future, whereas the epoch of reversed events occurred in our past (see Fig. 10.2). This is in contrast to Gold's suggestion, in which the backward-directed epoch lies in our *future*. This makes a big difference, because in Boltzmann's case causal influences always propagate away from, rather than into, the reversed-time region, thus avoiding entanglement horrors of the sort I discussed above.

For the same reason, Fred Hoyle and Jayant Narlikar were able to get away with a type of time-reversing cosmos, but one which didn't have to wait, à la Boltzmann, for absurd lengths of time in order for chance alone to do the trick.[6] In the Hoyle-Narlikar model, the universe contracts for an infinite duration, reaches a state of maximum compression, reverses (bounces), and expands out again forever (see Fig. 10.3). These scientists imaginatively suggested that the arrow of time always points away from the bounce point. We are located in the expanding phase, and the arrow of time points to the future, but in the contracting phase it points to the past. Of course, the situation is perfectly symmetric, so

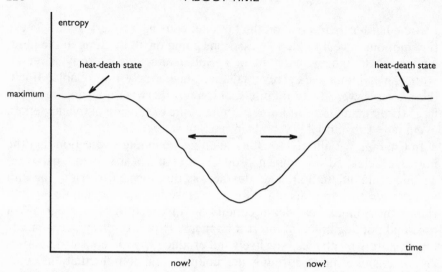

10.2 *The self-winding universe. The entropy of the universe remains close to its maximum value for eons. Very occasionally, it will undergo a large chance fluctuation, producing a dip in the graph. The situation is time-symmetric, but, unlike in Gold's model (depicted in Fig. 10.1), the arrow of time points away from the symmetry point.*

words like "expanding" and "future" can be interchanged with "contracting" and "past." Any sentient beings will perceive the universe to be expanding and time "going forwards," whichever epoch they inhabit. Causality precludes any communication between beings in oppositely directed phases. Because any antiworld aliens are located far in our past, they cannot know our future, so the problems that arise with contemporaneous time-reversed beings are avoided. Likewise, no wrecking influences can intrude from the antiworld into ours, because all such influences start out in our remote past and travel *backwards* in time relative to us, away from our epoch.

Unfortunately, the same cannot be said of Gold's theory, where the arrows clash head-to-head (Fig. 10.1). In this case, the past-directed arrow lies in our future, and influences from our epoch propagate causally forwards in time towards the antiworld. The same is true in reverse: antiworld influences come at us from the future, backwards in time. When these threatening influences arrive, trouble starts, as oppositely directed physical processes muddle together. Opinions differ on whether the resulting mess renders the theory totally invalid, or whether a sufficiently contrived setup could yet patch it all together nicely.

To get some idea of what needs to be patched up, consider the basic point of Gold's theory—that the flow of radiation from stars reverses when the expansion of the universe reverses. Although thermodynamics

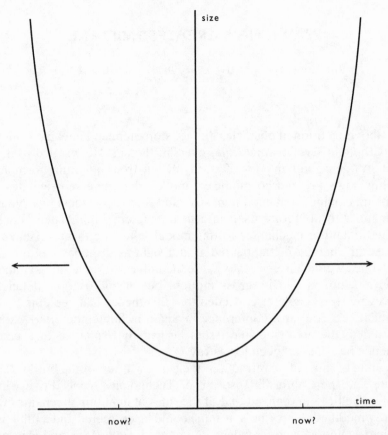

10.3 In this model by Hoyle and Narlikar, the universe contracts from an infinite size, bounces at a minimum, and expands out again forever. Time runs forwards in the expanding phase and backwards in the contracting phase, as judged by us (see arrows). The situation is, of course, perfectly time-symmetric.

lends the Gold conjecture a certain plausibility, it doesn't stand up well to scrutiny. The problem concerns the lag that exists between the expansion of the universe and the flow of heat and light through space. If the universe started to contract tomorrow, it would be billions of years before we began to see the most distant galaxies approaching rather than receding, because it takes light so long to traverse the universe. In order for the heat flow from the sun to start running backwards on cue, radiation from the depths of space would have to start converging on the sun billions of years before the expansion turned into a contraction. This requires that the universe is subject to a gigantic inbuilt conspiracy that can somehow anticipate future events in minute detail, which is a bit hard to swallow, but is maybe not impossible.

HAWKING'S GREATEST MISTAKE

He who never made a mistake never made a discovery.

SAMUEL SMILES

The above-mentioned obstacles did not stop Stephen Hawking from toy-ing with Gold-style cosmic time reversal, though. He was led to it, not from studying the behavior of starlight, but from quantum cosmology. In Hawking's favored cosmological model, the universe originates in a big bang, expands to a maximum size, and then recontracts in a symmet-ric manner to obliterate itself in a final "big crunch." When Hawking applied quantum mechanics to this model, after the fashion I have de-scribed in Chapter 7, it appeared at first sight as if the laws of quantum mechanics would automatically force the universe into a time-symmetric evolution, not just in its gross motion, but in microscopic detail too. However, Hawking later conceded that this theory had been his "great-est mistake," and at a conference in Spain in September 1991 entirely devoted to the arrow-of-time issue, he gamely explained to a packed audience how he had been led astray.[7]

In spite of this public retraction, the genie was out of the bottle. James Hartle and the California Institute of Technology Nobel Prize winner Murray Gell-Mann realized that, if the rules of quantum mechanics were slightly modified, Hawking's mistake could be corrected and a fully time-symmetric universe could indeed be imposed. Gell-Mann and Hartle did not propose that the universe *had* to be this way, only that it might be. There followed a lively but rather inconclusive discussion about whether, if they were right, we would notice anything unusual. Could we have any inkling at the present epoch of the existence of an antiworld in our far future? Gell-Mann and Hartle pointed out that it might be possible to spot some apparently irreversible processes gradually slow-ing down in anticipation of a reversal to come. The half-lives of some very long-lived radioactive isotopes, for instance, might perhaps sense the "turning of the tide" ten billion years down the road. The flow of radiation away into space might even now be very slightly inhibited. Perhaps the time was ripe for Partridge to repeat his experiment.

A TIME FOR EVERYBODY

A time to be born, and a time to die; a time to sow, and a time to reap.

ECCLESIASTES, 3:2

Meanwhile, Sydney philosopher Huw Price has attacked the physics community for "double standards," insisting that a universe which expands and contracts symmetrically *must* have an arrow of time that flips with it, on the basis that we have no right to distinguish one temporal extremity ("the beginning," or big bang) from the other ("the end," or big crunch).[8] Whatever physical or philosophical arguments we may invoke to make the arrow of time point away from the big bang towards the future, that same argument can be invoked to make it point away from the big crunch towards the past. Price's point is that, as the laws of physics do not distinguish one direction of time from another (ignoring kaons), and as the universe as a whole expands and contracts symmetrically, there is nothing physical to flag "start" and "finish."

The quantum cosmology of Hawking, Hartle and Gell-Mann, however, has an escape clause that enables it to evade Price's trap. To explain what it is, I have to recapitulate some pertinent facts about quantum mechanics. Recall from Chapter 7 that all quantum systems are subject to inherent uncertainty. As a typical system evolves, there are many possible outcomes, many contending realities, on offer. For instance, in the various laser experiments I described, a photon had a choice of which path to follow through a piece of apparatus. In the case of a lab experiment, the observer will always see a single, specific, concrete reality selected from among the phantom contenders. Thus a measurement of the photon's path will always yield a result of *either* one path or the other, but never both. When it comes to the universe as a whole, there is no outside observer, because the universe is all that there is, so quantum cosmology runs up against a major problem of interpretation. The favored way out is to assume that *all* the contending quantum realities enjoy equal status. They are not mere "phantom worlds," or "potential realities," but "really real"—all of them. Each reality corresponds to an entire universe, complete with its own space and time. These many universes are not connected *through* space and time, but are somehow "parallel," existing alongside each other. In general, there will be an infinity of them.

The existence of an infinite collection of universes and an infinity of times means that anything that is allowed to happen within the wide scope of quantum fuzziness *does* happen in at least one of the universes. Such a rich mosaic of existence enables the theory to have its cake and eat it. The overall quantum-mechanical evolution of the entire assemblage of universes is time-symmetric: it does not distinguish big bang from big crunch. However, each individual universe generally does possess a well-defined arrow of time. Thus there will be universes where this arrow points "forwards," and others where it points "backwards." Neither direction is favored. There will also be a very, very tiny fraction that undergo Gold-style reversals partway through. But a random observer is overwhelmingly likely to find her/himself in a universe with an unwavering arrow of time, and to define the past singularity, relative to this arrow, to be the big bang (origin) and the future to be the big crunch (end) of the universe. Bang and crunch will be physically distinct in the vast majority of cases.

You might be wondering why, if there are so many universes about, we see only one of them. This is explained by supposing that, when a universe splits into, say, two alternative worlds, observers also split, with each copy perceiving her or his single respective world. In practice, quantum processes going on at the atomic level all the time are continually splitting the universe, and the reader, into vast numbers of copies. Each version of you touchingly believes she or he is unique. Bizarre though this may seem, it is entirely consistent with experience, so long as the various universes remain separate. Trouble comes, however, if they start overlapping or interfering with each other.

This leads to a second question: is it possible to observe the other universes? The usual answer is no, but there is no unanimous agreement about this. David Deutsch, a physicist at Oxford University with a bent for the unusual, believes that microscopic experiments could in principle be done in which two or more worlds are temporarily joined, allowing physical influences to slip through.

What would happen if our universe (one of them!) became temporarily connected with one of the antiworlds? Would we be able to glimpse the future in a shadowy sort of way? Might we find objects in our universe doing apparently miraculous things (snooker balls coming together) because their time direction was temporarily reversed? Would we expect astonishing coincidences or happenings against the odds which, viewed in reversed time, are perfectly normal (such as a hand of cards being shuffled into suit order)? Alas, such is the stuff of science fiction. But science fiction sometimes gives a pointer to science fact, as the next chapter reveals. Under normal circumstances, a join between two quantum worlds would produce only atomic-level effects, rather than the sort of paranormal phenomena just described. However, some scientists suspect there *may* be circumstances in which a mixing of quantum realities dramatically manifests itself on a human scale.

CHAPTER 11
TIME TRAVEL: FACT OR FANTASY?

The problem here involved disturbed me already at the time of
the building up of the general theory of relativity.

ALBERT EINSTEIN

I do not take time travel very seriously.

ARTHUR C. CLARKE

SIGNALING THE PAST

Like many people, I first read H. G. Wells's story *The Time Machine* as
a teenager, and it left a lasting impression on me. In fact, it probably
contributed to my determination to become a scientist. The mark of a
great work of fiction is whether it can stand the test of time. *The Time
Machine* certainly falls into this category, and can be read with enjoy-
ment today even though it was published in 1895. Although this was a
full decade before the special theory of relativity, Wells anticipated some
aspects of Einstein's time with uncanny accuracy.

Several times I have made the point that, before Einstein, scientists
and philosophers generally thought of time as simply *there*. Physics was
about the behavior of matter and energy *in* space and time. The idea of
manipulating time didn't make much sense. Wells, however, conjectured that a machine employing physical forces could change time; in

particular, that the machine, together with any occupant, could travel through time in the way some machines can travel through space.

The theory of relativity brought time firmly within the scope of physics, linking space and time to physical forces and matter in a mathematically precise way. From the outset, it was clear that relativity permitted a type of time travel. The time-dilation effect, which I have discussed at length in earlier chapters, involves travel into the future. Remember the adventures of the twins Ann and Betty? When Betty zooms off to a star, she returns to find Ann older than she is. In effect, Betty has traveled into Ann's future. In principle, by traveling very close to the speed of light relative to Earth, Betty could return home in the far future, after millions of years have elapsed on Earth and all trace of humanity has vanished. Gravity can also slow time, enabling Betty to travel into Ann's future by spending time in a stronger gravitational field. In fact, we are all unwitting time travelers to a limited extent, because of the Earth's gravity. In this sense, then, time travel is a reality, and can be observed by the curious in any well-equipped physics laboratory. However, the really interesting question is, can a time traveler who goes into the future ever "get back" again? It's all very well having the possibility of using high-speed motion or intense gravitational fields to reach the far future, but if you are then stranded there, the appeal of time travel is somewhat tarnished.

Coming back from the future amounts to traveling into the past, and on that score the theory of relativity is far more obscure in its predictions. Before getting into that, let me emphasize the need to keep a clear distinction between time *reversal*, of the sort discussed in the previous chapter, and time travel. In the former case, the arrow of time itself is inverted, making time "run backwards." Travel into the past, by contrast, leaves the direction of time unchanged, but involves somehow visiting an earlier epoch.

In Chapter 3, I discussed the subject of tachyons—hypothetical particles that would always travel faster than light—and mentioned that "faster than light" can mean "backwards in time." Let me now explain why this is so. Suppose we have a gun that can shoot particles at a target. First consider the case of ordinary bullets. Experience and common sense indicate that the bullet strikes the target *after* it is fired. If we call the act of firing the gun event E_1, and the arrival of the bullet at the target event E_2, then we can be quite sure that the time sequence of these two events is E_1E_2. Now, the theory of relativity predicts that the duration of time between E_1 and E_2 can vary according to the state of motion (or the gravitational situation) of the observer. However, the theory also makes it clear that, however much the interval E_1E_2 is stretched or shrunk, the time *order* E_1E_2 is never reversed. In other words, the before-after relationship is unaffected by motion or gravity, even though the *duration* might be.

All this changes when tachyons are allowed. If the bullet was tachyonic, and sped to its target faster than light, then it is possible for an observer to see the bullet hit the target *before* the gun is fired! For example, suppose the bullet travels at twice the speed of light; then someone moving in the same direction as the bullet at 90 percent of the speed of light would see the target shatter first and the gun fire afterwards. The bullet would appear to travel backwards from the target into the barrel of the gun. Someone traveling at half the speed of light in the same direction would see the bullet travel at infinite speed, leaping from gun to target instantaneously. For faster-than-light motion, the time sequence of events E_1E_2 is no longer fixed, but can appear reversed, as E_2E_1, in certain reference frames. In those frames, the tachyons appear to travel backwards in time relative to normal physical processes.

Tachyonic motion falls short of H. G. Wells's dream, because it does not allow normal matter, of the sort that people are made of, to travel into the past. However, if tachyons exist and can be freely manipulated, this would at least enable us to send signals into the past, even if we couldn't travel there. This is how Ann and Betty could do it. Betty is out in space speeding towards her favorite star at 80 percent of the speed of light. At noon on Earth, Ann beams a tachyonic signal to Betty at four times the speed of light, relative to the transmitter on Earth. As far as Ann is concerned, the signal reaches Betty some time later. But Betty sees things differently. From her point of view, the signal arrives *before* Ann sent it. (Some people might argue that, from Betty's reference frame, it is Betty who sends the signal to Ann, but I shall not dwell on the semantic aspects of this tricky scenario.) The next step is for Betty to reply, using a tachyon transmitter as well. Suppose Betty's tachyons also travel at four times the speed of light relative to the transmitter, but this time the transmitter is in the speeding rocket. It is now Betty who reckons the tachyons will arrive after they are sent, whereas Ann, back on Earth, receives them before Betty transmitted them. As far as Ann is concerned, the outgoing signal travels into the future, and the reply travels into the past. By a suitable arrangement of speeds, the reply can get back to Earth before the original signal is sent. This startling possibility was well understood by Einstein, who clearly took a dim view of it. In his 1905 paper, he wrote that velocities greater than light "have no possibilities of existence." It was a sentiment widely echoed among his colleagues. "The limit to the velocity of signals," wrote Eddington, "is our bulwark against the topsy-turvydom of past and future. . . . The consequences of being able to transmit messages concerning events Here-Now [faster than light] are too bizarre to contemplate."[1]

For those readers who like to put numbers in, here is an explicit example. Suppose Betty leaves at 10:00 A.M. and at noon (Earth time) Ann sends the original signal towards Betty at four times the speed of light relative to Earth. Traveling at 80 percent of the speed of light,

Betty will receive the signal at 12:30 P.M, Earth time, by which time she will be two light-hours out in space, as observed in Ann's reference frame. Betty's clock will, of course, indicate something quite different. The 2½-hour journey from Earth will seem to her like only 1½ hours, because of the time-dilation factor of 0.6. The rocket clock will therefore indicate 11:30 A.M. The distance from Earth as measured in Betty's reference frame will also be different. As far as Betty is concerned, it is Earth that is receding, at 0.8 times the speed of light, so her 1½-hour journey will have put a distance of $0.8 \times 1.5 = 1.2$ light-hours between her and Ann. If Betty replies without delay, her signal, traveling at four times the speed of light this time in *Betty's* reference frame, will complete the return trip in ⅜ hour, or 22½ minutes, Betty time, arriving on Earth at 11:52½ A.M. as viewed from Betty's frame of reference—i.e., 1⅞ hours after her departure, by which time she is 1½ light-hours from Earth in her frame. But in Betty's frame, it is *Ann's* clock that is dilated, by a factor of 0.6. The 1⅞-hours' total journey time as told on Betty's clock therefore translates into $1⅞ \times 0.6 = 1⅛$ hours on Ann's clock, which therefore reads 11:07½ A.M. This means that the return signal arrives on Earth 52½ minutes *before* Ann sent the original signal.

VISITING THE PAST

Time is nature's way to keep everything from happening at once.

JOHN WHEELER

Although Einstein's special theory of relativity unequivocally forbids ordinary matter, and by implication human beings, from traveling into the past, the general theory of relativity is flimsier on the subject. Shortly after the theory was published, Hermann Weyl noted that in a spacetime with a particular gravitational arrangement a person's world line—his or her path in spacetime—might loop back and intersect itself. Weyl's point is that, even though *locally* a particle may never exceed the speed of light, *globally* its future might connect up with its past. This possibility arises because a gravitational field implies that spacetime is curved, and the curvature might be great enough and extended enough to join a spacetime to itself in novel ways. To see what I have in mind, look at Fig. 11.1. Here spacetime is bent into a loop in two different ways. In (a) space is bent around to connect up with itself. If the universe had this geometry, an observer could travel around the universe and return to her starting point. In (b) spacetime is bent in the time direction instead and connects up with itself in the past. In this arrangement, an observer

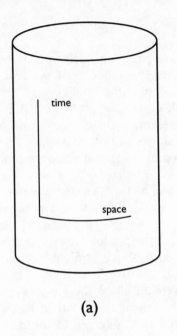

(a)

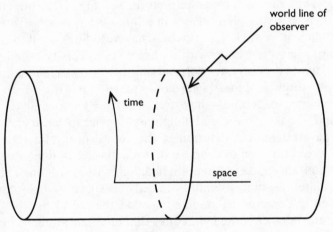

(b)

11.1 Bending spacetime into loops. Shown is one dimension of space plus time. In (a) space is bent into a loop, implying it is finite in extent, so the universe could be circumnavigated. In (b) time is bent into a loop. The world line of a static observer may eventually "circumnavigate time" and join up with itself.

who simply remains at rest in space will eventually return to an earlier point in time.

The distinction between going back in time by traveling faster than light, and by warping the spacetime itself, is a crucial one, best illustrated by introducing the concept of the "light cone." In Chapter 2, I explained the concept of a Minkowski diagram (Fig. 2.2), in which space and time are represented in the same picture. I now wish to elaborate this somewhat. Figure 11.2 (a) shows a Minkowski diagram with two space dimensions left out. Time is drawn vertically and space horizontally. The world line of a typical object, labeled "A," is also shown. The refinement I have introduced into this picture is to include the spacetime paths of two light pulses (i.e., the world lines of two photons), emitted from A at some instant E and flying off through space. You can envisage a momentary flash of light at E. One photon travels to the right, the other to the left, and they trace out oblique straight-line paths in spacetime. The paths must be straight, because light always travels at the same speed. If the unit of measurement along the space axis is chosen to be light-years, and the unit of time is years, then the world lines of the two photons will lie at forty-five degrees in the picture. It is easy to add another space dimension to the picture; see Fig. 11.2 (b). The flash of light at E now sends photons out in all directions in a horizontal plane, rather than just to the left and right. The world lines of all these photons lie along an inverted cone with its apex at E. This is called the "light cone." There is no need to worry about the third space dimension.

We may draw imaginary light cones starting out (i.e., with the apex) at any event in spacetime—in particular, at any point along the particle's world line. Because the speed of light is a barrier to cause and effect, the light-cone arrangement determines the causal properties of the spacetime. There is no need to suppose that real flashes of light are emitted to analyze causality: a set of hypothetical light cones will do. The rule that normal material objects cannot exceed the speed of light can now be conveniently depicted by demanding that the world line of the object always remains *within* the light cones that originate along it. Fig. 11.3 (a) shows the world line of a particle that moves about, and a series of light cones enveloping it. The world line dutifully avoids sloping over far enough to penetrate any of the cones. By contrast, Fig. 11.3 (b) shows the unlawful behavior of a particle that accelerates through the light barrier, and pierces one of its light cones. This will happen when the slope of the particle's world line is greater than forty-five degrees, indicating it is moving faster than light. So there is a fundamental rule of relativity: the world lines of ordinary objects are not allowed to pierce any of their light cones.

These pictures make it clear why going faster than light can take you back in time: if the world line of a particle were to spear the light cone

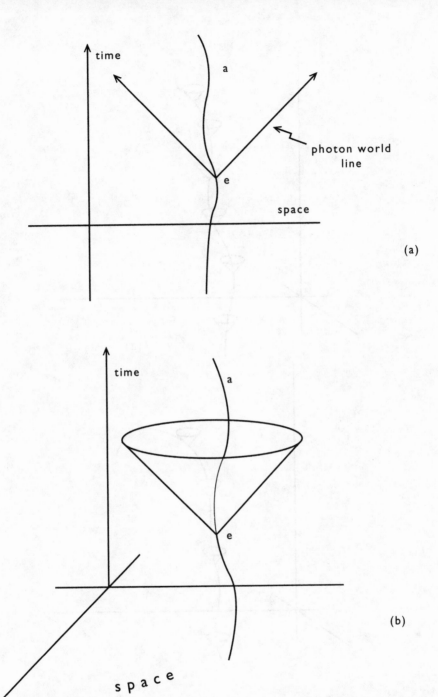

11.2 *The light cone. At an instant e the body a emits a flash of light. Photons travel out in all directions at a fixed speed. In (a), where only one dimension of space is drawn, the photons' world lines reduce to two oblique lines, representing left- and right-moving photons respectively. In (b), two space dimensions are shown, and the photons' world lines fill out the surface of an inverted cone—the light cone.*

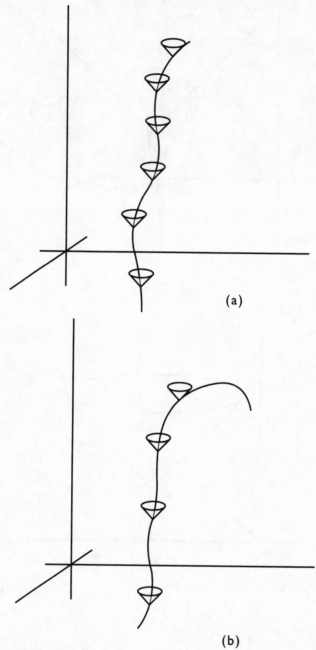

(a)

(b)

11.3 Through the light barrier! (a) The particle's world line always remains within its light cones, in accordance with the theory of relativity. (b) The world line starts out well behaved, but then curves over sharply, indicating a rapid acceleration to superluminal speeds. As a result, the world line pierces a light cone. It may even curve over into the past (broken line). According to the theory of relativity, such behavior is impossible.

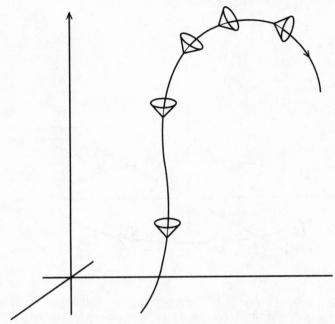

11.4 Tipping into the past. Gravity affects light, and may tip the light cones over so far that the world line of a particle "bends over backwards," taking it into the past. Note that time as told by the particle also tips over and "runs the wrong way" (arrow) relative to the time coordinate shown (which represents the frame of an observer far from the gravitational field).

and then curve right on over, it could loop back in time and connect up with a region in its own past. Since we are discounting this possibility, let us move on to the second and more plausible scenario. I have been at pains to explain how gravitation is a distortion of spacetime geometry. If spacetime is warped, the light cones will be warped too. A gravitational field may then have the effect of tipping the cones to one side (and maybe stretching or shrinking them too, but I shall neglect this). If the cones tip, the world lines of material objects must tip with them, because they are not allowed to pierce the light cones under any circumstances. It may happen that the cones tip right over on their sides: this is the case at the surface of a black hole, for example.

Figure 11.4 shows a succession of light cones that gradually tip over, allowing the world line of a particle within them to curve too—so far, in fact, that it bends downwards, in the direction of the past. If spacetime was really like this, the world line could loop back and intersect itself, which amounts physically to the object visiting its past. The world line might even connect up with itself to form a closed loop, in which case the object becomes its own past self. A similar situation with tipped light

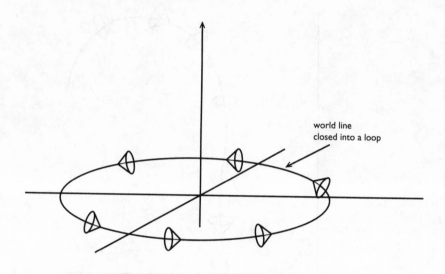

11.5 Gödel's solution. Kurt Gödel discovered that, if the universe rotates, the light cones can tip on their sides in the manner shown, permitting a particle's world line to form a closed loop.

cones is depicted in Fig. 11.5, where again a world line forms a closed loop in the horizontal plane.

The important point to note about these light-cone patterns is they permit time travel to be achieved without the material body concerned anywhere exceeding the speed of light. Locally, the world line always remains within the nearby light cones and the rules of special relativity hold; globally, however, the light-cone structure is so distorted that it permits world lines to intersect themselves. In such a scenario, the route to the past lies in going around some sort of spacetime loop; it does not consist of staying put and "going in reverse" through previous events, as H. G. Wells intimated. We can imagine a hypothetical time traveler going off, Betty-like, on a round-trip in space and returning to Earth to find it at an earlier epoch than the moment of departure.

Extravagant though the idea of self-intersecting world lines may seem, such a possibility lies buried within Einstein's general theory of relativity. The first explicit example containing time loops was provided by the Austrian mathematician Kurt Gödel, an eccentric and reclusive logician who worked alongside Einstein at the Institute for Advanced Study in Princeton. In 1949, Gödel published a novel solution of Einstein's gravitational-field equations describing a light-cone structure very similar to that of Fig. 11.5. The Gödel solution is not very realistic, because it assumes that the whole universe is rotating, something pretty well ruled out by observation. Nevertheless, it served to demonstrate that there is nothing intrinsic to the theory of relativity that forbids a particle of matter, or in principle a human being, from reaching the past—and going back to the future. Gödel himself wrote of his solution: "By making a round trip on a rocket ship in a sufficiently wide course, it is possible . . . to travel into any region of the past, present and future, and back again."[2]

Following the publication of Gödel's solution, Einstein confessed that the prospect of a spacetime geometry that permitted time loops had troubled him from the very start, even during his early formulation of the general theory.[3] He noted the physical problems and causal paradoxes that came with this possibility, but left open whether solutions such as Gödel's should always be discarded on physical grounds.

BLACK-HOLE TIME MACHINES

By this stage of his career, Einstein had largely abandoned mainstream physics. He spent the war years isolated and working quietly on his own theories. As a foreign Jew, confirmed pacifist and independent-minded intellectual who espoused a variety of political causes, he was regarded with suspicion by the security forces. Certainly he was not suited to work on the atomic-bomb project. At the end of the war, he officially retired, although he kept an office at the Institute in Princeton, and divided his time between there and his home. Although he would occasionally attend a seminar, and continued to read the professional journals, he contributed little to the exciting developments in subnuclear particle physics and quantum field theory that were sweeping through the physics community in the postwar years. He retained just one or two collaborators, and worked obsessively on attempts to construct a unified field theory which would unite his magnificent theory of relativity with quantum physics in a way that was not philosophically objectionable to him. He never succeeded.

Because Einstein died before the full renaissance of his general theory of relativity, he had little to say about the modern developments that lead to outlandish ideas like black holes and time travel. In spite of the power and beauty of his general theory, it was relegated to the backwaters of physics for decades, largely because its predicted effects were usually small and hard to test. Gravitation theory was the preserve of a very few specialists, mainly those with astronomical or cosmological interests. But postwar developments changed all this. Radio astronomy opened up an additional window on the universe. The era of artificial satellites provided an opportunity to view the universe at wavelengths inaccessible from the ground, while improvements in ground-based telescopes and the increasing use of electronic computers transformed astronomers' ability to map the universe in great detail.

Observational advances were matched by a resurgence of interest in theoretical questions. The possibility of gravitational waves began to be taken seriously. The need to combine gravitational physics with quantum mechanics inspired Wheeler and his collaborators to investigate strong gravitational fields, gravitational collapse, and spacetime topology. New mathematical techniques were developed, textbooks were written, and the subject of general relativity finally blossomed into a fully developed discipline a decade or so after Einstein's death.

It was purely as an accidental by-product of these developments in gravitation theory that another possibility for time travel was uncovered, by a New Zealand mathematician named Roy Kerr.[4] It came from the study of black holes. Schwarzschild's solution had served well for many decades, but it was clearly unrealistic in one respect. A real star would undoubtedly be spinning when it collapsed, and nobody knew the solution to Einstein's equations corresponding to a rotating black hole, until Kerr found it.

Outside the event horizon, the properties of Kerr's black-hole solution are broadly similar to those of the Schwarzschild case, but the interior is structured quite differently. Whereas a particle (not a tachyon) falling into a Schwarzschild hole necessarily hits the central singularity after a short time, a particle falling into a Kerr hole may avoid the singularity altogether. Where does it go? Nobody really knows, but Kerr's solution gives a possible answer. Just as Schwarzschild's solution can be extended into an antiworld—another universe, where time runs backwards—so Kerr's solution extends into an infinity of other universes, both worlds and antiworlds! In addition, there is a weird region *within* the black hole where the light cones tip over, Gödel-like, and permit world lines to form closed loops.

Unfortunately for aspiring temponauts, most experts think the Kerr solution would not apply throughout the interior of a real black hole. Quite apart from its idealized mathematical form, which may or may not resemble a real spinning black hole, the gateway to the other universes and the time-travel region described by this solution turns out to be intrinsically unstable. Furthermore, the singularity inside the Kerr hole is "naked." This means it can be seen by an observer in the interior region. By contrast, the singularity inside the Schwarzschild hole lies to the future of all observers: they don't know it is there until they hit it. A naked singularity is far more fearsome. Remember, singularities are edges or boundaries where space and time cease to exist. Because the laws of physics break down there, it is impossible to know what might come out of a singularity. So long as the singularity remains hidden, we need not worry too much about it, but a naked singularity could influence events in an unknown way, making attempts to draw any physical conclusions hazardous. (For an in-depth discussion of the implications of naked singularities, see my book *The Edge of Infinity*.)

In 1974, the physicist Frank Tipler discovered yet another time-travel solution to Einstein's equations involving rotation, this time of a cylinder of matter.[5] The time-travel region lies close to the surface of the cylinder. Because there are no singularities in the Tipler model, it appears slightly more physical than the Kerr example. However, it is not without its problems. Most notably, the solution describes an infinitely long cylinder —an obvious fiction. Moreover, for time loops to occur, the cylinder has to spin on its axis at a fantastic speed, and risks flying apart from centrifugal forces unless it is composed of material vastly more dense than nuclear matter. All in all, it is far from clear that time loops would occur in the case of a more realistic spinning object.

WORMHOLES AND STRINGS

None of the aforementioned obstacles have prevented science-fiction writers from enthusiastically embracing the subject of time travel. Following Wells's trail-blazing novel, many writers have speculated about creating time machines using unusual states of matter or gravitational fields. It was actually a work of fiction that triggered the only systematic investigation of time travel in the history of science. In a fast-moving story called *Contact*, Carl Sagan relates a saga of alien beings who send a radio message to Earth containing details for the construction of a wonderful machine.[6] The scientists who build the machine are able to travel to the center of the galaxy very rapidly. They achieve this not by going faster than light but by traveling through a so-called wormhole in space.

The term "wormhole" was also coined by John Wheeler, he of the black hole. In the 1950s, Wheeler envisaged the possibility that two points in space might be connectable by more than one route. The original idea is shown in Fig. 11.6, which represents space in terms of a two-dimensional sheet. A and B are two points in space. To get from A to B by normal means, you would follow the dotted path. But there might exist a tunnel or tube (the wormhole) providing an alternative route (the dashed line). The possibility of *two* routes connecting the same points in space is another example of how, in general relativity, spacetime may be bent around far enough to reconnect with itself, thus providing the possibility of loops in both space and time.

As shown, the wormhole route looks longer, but this is because my drawing is largely symbolic. (Remember that spacetime diagrams can greatly distort distances.) A careful mathematical study shows that

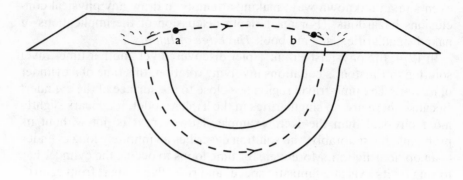

11.6 Wormhole in space. Two points, a and b, may be connected by a path through "normal" space (dotted line) or one that threads through the wormhole (dashed line). The length of these two routes may be very different. As drawn, the wormhole path looks longer, but under some circumstances it can be much shorter.

wormholes can actually *shortcut* the distance from A to B. This is made more plausible if the diagram is folded over, as shown in Fig. 11.7, where the wormhole now appears as a short tube. Remarkably, Einstein anticipated this type of geometry in work carried out with Nathan Rosen in the mid-1930s. For that reason, a wormhole is sometimes known as an "Einstein-Rosen bridge." It may happen that an astronaut can get from A to B through the wormhole *faster* than light can get there via the "normal" route. By outpacing light in this manner, the astronaut can also travel backwards in time.

If we want to turn the wormhole into a time machine, the wormhole mouths have to be regarded somewhat like our twins Ann and Betty. One end (Betty) needs to be accelerated (somehow!) close to the speed of light relative to the other end (Ann), stopped, and accelerated back again. The itinerant mouth would then return some time in the future of the static mouth, and a permanent time difference would be established between the two ends of the wormhole. To travel into the past, an astronaut has to pass through the wormhole in the right direction, and then zoom back to her starting point through "ordinary" space at high speed, forming a closed loop in space. If the circumstances are right, her world line will also form a closed loop in time.

To investigate whether this fantastic state of affairs is physically possible, or just a wacky speculation, Kip Thorne and his co-workers at Caltech embarked on an extended program of research.[7] They were guided in their task by the fact that a form of wormhole was already known in connection with black holes. As I mentioned in Chapter 10, an

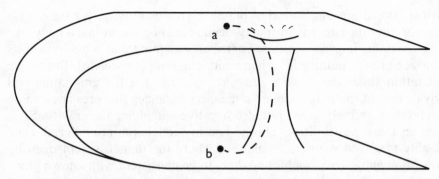

11.7 Shortcut through space. If the space in Fig. 11.6 is folded over, the wormhole straightens out and is more suggestive of a shorter route between a and b.

idealized extension of the Schwarzschild black hole links our universe to another, in which time runs backwards. Spatially, the bridge linking the two universes has the form of a tunnel or wormhole like that shown in Fig. 11.7. However, remember that it isn't possible for a particle of ordinary matter to traverse the Schwarzschild wormhole and emerge in the other universe. To do so, it would need to travel faster than light.

The reason for the restriction on traversing the Schwarzschild wormhole is that in this model the wormhole isn't static, as drawn, but in effect briefly winks open just once and then closes off again. It does this so rapidly that there simply isn't enough time for a particle (or an astronaut) to get through before it pinches shut and squashes everything inside to oblivion. Thorne's team cleverly conceived of a way of keeping the throat open long enough for something to get through. In their case, however, they had in mind a wormhole that connects a region of space, not to another universe, but (following Sagan) to a different part of our own universe, as depicted in Fig. 11.7.

To shore up the wormhole against its natural tendency to collapse, Thorne invoked another of H. G. Wells's favorites—antigravity. Physically, antigravity means some sort of substance or field that gravitationally repels instead of attracting. All ordinary states of matter are attractive, so the researchers had to appeal to some sort of exotic (one might be tempted to say "quixotic") state. They were naturally drawn to quantum physics, that cornucopia of exotica. Buried obscurely in the physics literature were a number of mathematical studies indicating how certain peculiar quantum states can, in a very limited way, produce antigravity. Sometimes this occurs because the energy of the quantum field can go negative here and there. Energy implies mass, so negative energy spells negative mass, and antigravity—in theory.

Another possibility for quantum antigravity also presented itself. In general relativity, *pressure* as well as mass is a source of gravitation.

Most people are unaware that pressure creates gravity, for the simple reason that its effect is normally negligible. For an ordinary body like the Earth, its internal pressure contributes no more than a billionth of its surface gravity (adding less than a milligram to your weight). But exotic quantum states can have a pressure so great that the gravitating pull rivals that of the mass. Under some circumstances this pressure can be not only extremely large, but also negative—implying antigravity. Seizing on these possibilities, Thorne and his collaborators analyzed some highly idealized wormhole solutions where the tunnel is kept open by quantum antigravity, and found them to be consistent with known physics. Their initial research has led to a flood of papers on the subject, and its ramifications continue to be explored.

Meanwhile, an entirely different sort of time machine has been proposed by Richard Gott of Princeton University, using objects known as "cosmic strings."[8] Cosmologists have toyed with the idea that just after the big bang, when the universe was exceedingly hot and dense, the various quantum fields present tied themselves into knots and twists in such a way as to produce extremely narrow threadlike tubes of concentrated field energy. These tubes, or cosmic strings, cannot easily "unwind" themselves, and would be frozen as relics, perhaps until the present day. Astronomical searches for cosmic strings have so far proved inconclusive.

The gravitational properties of cosmic strings are very odd. A loop of string gravitates much like any other body, but a straight segment exerts no direct gravitational force, even though each kilometer may contain as much mass as the Earth. However, straight strings still affect light, so they have implications for the causal structure of spacetime. Gott found that, if two infinitely long parallel cosmic strings fly apart at high speed, the light cones will tip far enough to allow particle world lines to loop back into the past. Of course, Gott's scenario is pure make-believe. It suffers from the problem that the strings have to be obligingly formed with the required configuration and motion, as well as other physical difficulties associated with their infinite extension.

The upshot of this recent flurry of work on time travel is that there is nothing very obvious in the laws of physics to prevent it happening *in principle*, although in all the examples studied time loops can be achieved only by manipulating matter and energy in the most extreme and fanciful manner. Nevertheless, let's accept for the moment that a time machine could, in principle, be constructed (or discovered in nature). What would be the consequences?

PARADOX

Time is clothed in a different garment for each role it plays in our thinking.

JOHN WHEELER

Anyone who has critically read *The Time Machine* or watched the movie *Back to the Future* will have spotted that travel into the past, or even the possibility of signaling the past, opens up a Pandora's box of puzzles and paradoxes. The most famous of these is called the grandmother (or grandfather) paradox. Suppose a time traveler were to go back into the past and murder his grandmother. As a result, the said traveler would never have been born. But then he could not have carried out the murder after all, in which case he *would* have been born. . . . Either way, there's a mind-jolting contradiction.

The paradox erupts because the present state of the world is *determined* by the past. Altering the past is therefore likely to lead to trouble, because it can have escalating knock-on effects that uncontrollably and inextricably weave themselves into the fabric of the present. Even a single subatomic particle sent backwards in time could change the present state of the world dramatically. The particle might be part of a coded signal, for example, that would trigger a major response in the receiver. Or it might divert the course of evolution. (A single encounter between a cosmic-ray particle and a molecule of DNA can cause a crucial mutation.) But what sense can be ascribed to a *changed* past, still less to a changed present? The present state of the world is what it is; it can't be turned into something else. The issues involved in these musings go beyond mere science fantasy. The laws of the universe must by definition describe a consistent reality. If time travel inevitably leads to irresolvable paradox, then it cannot be allowed within the framework of physical law. In this case, should we discover that our best current theories permit travel into the past, even in exceedingly contrived and unrealistic circumstances, then those theories would be suspect.

Paradox is circumvented if the causal loops are self-consistent. It would then be the case that the actions of the time traveler are already incorporated in the deterministic web that links past and present. The traveler who squashes a beetle and changes evolution will do so in a way that produces precisely the biological circumstances of the world from which he has come. But killing grandmothers is out. This would seem to place strong restrictions on free will, but there does not seem to be anything logically objectionable to the prospect of causal loops that thread past and future together consistently.

The grandmother paradox is actually but one of a set of problems deriving from the possibility of travel into the past. The time traveler might, for instance, encounter a previous copy of himself, in which case there would be two of him! This bewildering state of duplication could be achieved merely as a result of going back in time by one second. By repetition, an unlimited number of copies of the time traveler could be created at any one time (recall Feynman's zigzagging electron duplicating itself across the universe). Though there is no logical paradox involved here, the prospect of the unrestricted duplication of objects makes our minds reel, and plays havoc with some cherished laws of physics (such as the law of conservation of energy).

David Deutsch, the expert on the many-universes theory I mentioned in the last chapter, has made a careful study of time-travel puzzles and their possible solutions.[9] He has pointed out an even more vexing conundrum than the grandmother paradox, one which strikes at the very heart of scientific rationality. Consider the example of a time traveler from 1995 who visits the year 2000 and learns of a marvelous new solution of Einstein's equations, published in an edition of that year's journal *Physical Review* by an obscure scientist named Amanda Brainy. The traveler returns to his own year, armed with a copy of the solution, and seeks out the young Amanda, finding her to be a first-year physics student at his local university. He then sets out to teach her relativity, and eventually shows her the new solution, which she duly publishes under her own name in *Physical Review* in the year 2000. The problem about this little story is: Where did the knowledge of the new solution come from? Who made the discovery? Amanda didn't: she was told the solution by the time traveler. But he didn't make it either; he merely copied it down from her paper in the journal. Although the story is entirely self-consistent, it still leaves us feeling bemused and dissatisfied. Important new information about the world can't simply *create itself* in that manner, can it?

So horrendous are the physical and philosophical problems with time travel that Stephen Hawking has proposed a "chronology protection hypothesis," according to which nature will always find a way to prevent wormholes and other contrivances from permitting travel into the past.[10] That way, remarks Hawking, the universe can be made safe for historians. There is no general agreement about whether chronology protection is valid, and if so whether it is contained in existing physics or requires something new. All the known examples of time travel have pathological features that would render them unphysical or unstable in practice. But without a general theorem to rule out *all* loopy spacetimes, there is always a chance that some clever researcher will come up with a physically realistic example of how to beat the clock—and survive.

A shaky argument that is often used against time travel is that if our descendants ever discover how to do it they will come back and visit us.

As we don't see these temponauts, we can conclude that they will never come to exist. Stephen Hawking used this reasoning to beef up his chronology protection hypothesis, remarking that "we have not been invaded by hordes of tourists from the future." However, most of the time machines that have been mooted so far don't allow travel to a time prior to the construction of the machine, so if we were to build one today we couldn't go back and witness the Battle of Hastings, for example. Nor could our descendants use such a machine to visit us. Only if an ancient alien civilization were to make us a gift of an old time machine, or if nature had spontaneously created the necessary wormhole in the remote past, could we visit epochs before our own. The absence of travelers from the future can't therefore be used to rule out time travel completely.

The toughest argument against visiting the past is undoubtedly the grandmother paradox, and a lot of thought has gone into ways of avoiding it. One let-out is to invoke the many-universes idea, which I touched on in Chapter 10. If there exist many similar parallel worlds, it may be that a visit to the past takes you back to an earlier epoch, not of your own world, but of a closely similar quantum version. The murderous temponaut then finds he has killed a carbon-copy grandmother of one of his parallel selves, leaving his own future world untouched. This deft resolution assumes we can mix and match quantum worlds on a macroscopic scale, permitting the temponaut to step across into a parallel reality, and back again, having altered it significantly. It seems to me like a very fanciful extrapolation of the many-universes theory. In any case, the entire question of the behavior of quantum particles in a world where time loops are possible is still being actively researched.

The impact of time-travel paradoxes derives more from their dazzling psychological effect than from any logical quirkiness. Human beings are locked into thinking of time as something that flows like a river. It then seems puzzling how a time traveler can sail into the past. How can he glide along, always downstream, only to find himself upstream again, without scrambling out of the current and hitching a ride back along the bank? The prospect of riding the temporal current around in a closed loop delivers the same disorienting shock as an Escher painting. At the end of Wells's novel, the narrator speculates about the fate of the time traveler, who ominously failed to return from his latest trip: "He may even now—if I may use the phrase—be wandering on some plesiosaurus-haunted Oolitic coral reef, or beside the lonely saline seas of the Triassic Age."[11] The "now" here betrays the essential temporal double-think; it is as though our time somehow goes with the traveler in the machine, a sort of tributary of the twentieth-century river of time, winding back through the eons and mingling with the Triassic river of time. But that is absurd. The Triassic Age isn't *now*, it is *then*. Or is it?

CHAPTER 12
BUT WHAT TIME IS IT *NOW?*

Time, like an ever-rolling stream, bears all its sons away.

ISAAC WATTS

In fact, time, the ever-rolling stream, has no more to do with the existence of clocks than with that of sausages.

HERBERT DINGLE

CAN TIME REALLY FLOW?

No position is so absurd that a philosopher cannot be found to argue for it.

MICHAEL LOCKWOOD

There is a famous sketch in the British TV comedy series *Monty Python's Flying Circus* that pokes fun at Australian philosophers. Generally, philosophers are an easy target for humorists, perhaps because they often seem to be defending propositions that strike most people as manifestly ridiculous. Well, there has probably been more nonsense written by philosophers on the subject of time, from Plato onwards, than on any other topic. However, by common consent, one of the few philosophers to bring some sanity to speculations about the nature of time is Jack Smart, who lives up to his name and is as Australian a philosopher as they come. As he spent much of his working life at The University of Adelaide, I feel a certain affinity.

I first met Jack Smart when he visited Britain in the early 1980s and delivered a thought-provoking lecture on quantum physics and time at the University of Newcastle. His delivery was continually interrupted by a nitpicking scientist who had a fixation about the contention (based on quantum mechanics) that material objects aren't "really there." Jack is a polite and gentlemanly soul, but finally he lost his cool: "I wish *you* weren't there!" he expostulated, and the interruptions ceased.

Smart once wrote: "Talk of the flow of time or the advance of consciousness is a dangerous metaphor that must not be taken literally."[1] In other words, the "river" of time is not really there. That may seem as absurd as claiming that material objects are not really there, but Smart is on firmer ground on this one. I have already explained how the theory of relativity leads to the notion of block time, and the picture of time as the fourth dimension simply "laid out all at once." Since Einstein, physicists have generally rejected the notion that events "happen," as opposed to merely *exist* in the four-dimensional spacetime continuum.

It is not only physicists who have trouble making time pass. For decades, philosophers have sought to pin down this elusive flux, only to find it slip from their grasp in a linguistic muddle. Oceans (rivers?) of ink have been spilled on the subject, yet still the flow of time is as mysterious as ever. So mysterious, in fact, that philosophers such as Smart have been forced to conclude that there is no river of time. It is, so to speak, all in the mind. "Certainly we *feel* that time flows," Smart concedes, but in his opinion, "this feeling arises out of metaphysical confusion." In fact, he believes it is merely "an illusion."

What sort of illusion could it be? There is a rough analogy that helps. If you spin around quickly and then stop, the world "keeps going." The sensation of dizziness creates the impression that the universe is in motion—in a state of rotation, in fact. Of course you know it isn't. You only have to fix your gaze intently on the wall of the room to *see* it isn't. Intellectually, you can reason the motion away. Yet it still *feels* as if the world is moving. Perhaps the flux of time is just a feeling too, and when we fix the steely gaze of rationality on the events of the world it simply melts away.

The unreality of time's passage has been near the top of the philosophical agenda from the outset. Parmenides thought he had disposed of the time-flux concept over two millennia ago, when he reasoned that all change was impossible. His argument was simply that, since everything is what it is, and cannot be what it is not, then nothing can change from what it is into what it is not. Nothing can come out of nothing, and "being" is complete in itself. There are no half-measures, said Parmenides, no state of part-being, part-nonbeing through which an entity may slither on the road to *becoming*. Similar quirky reasoning was deployed by Zeno of Elea, who argued that all *motion* was impossible, since at any given instant of time an apparently moving object is in fact static.

Zeno considered the flight of an arrow and pointed out that at every moment in its trajectory it occupies one and only one "block" of space. Since it could not occupy more than one space at a particular moment, it must be stationary at that moment. And as this state of affairs is true at each and every moment, there can be no motion of any kind. The world is frozen!

This honorable tradition of reasoning time or change into nonexistence on philosophical grounds has continued into the modern era. About the turn of the century, the bespectacled Cambridge metaphysician and churchgoing atheist John McTaggart argued that the concept of time is so riddled with contradiction that it makes more sense to suppose that time does not exist at all.[2] Impressions of temporality, argued McTaggart, are mere human inventions. He arrived at the awesome conclusion that time is unreal by considering how the *moving* division of time into past, present and future is blatantly incompatible with the *fixed* dates that can be attached to events. The incompatibility between a moving now and a static time coordinate was brilliantly encapsulated in Jack Smart's arresting question: "How fast does time flow?" We all know the answer: one second per second. The muddle in the metaphor manifests itself immediately. Speed is defined as distance moved per unit of time. How can time move "in time"?

An entertaining writer named J. W. Dunne published a popular book in 1927 called *An Experiment with Time* in which he claimed to have solved the problem of how to make time flow.[3] He cunningly invoked a second time dimension to act as a measure against which to gauge the speed of the first. Unfortunately, there is no scientific evidence for more than one time, and Dunne's argument seems decidedly *ad hoc*. It also runs into the problem of how to time the second time dimension, for which Dunne was quite prepared to introduce a third, and then a fourth, and so on into the open jaws of an infinite regress.

But without a time to time time, how can time move? Smart reminds us of the metaphor of the river, "bearing us inexorably into the future toward the big waterfall which is our death." Alternatively, instead of our being carried along remorselessly atop this temporal current, we think of ourselves as spectators sitting on the riverbank witnessing the events of the future sweeping down upon us, and those of the present receding into the past. But Smart dismisses such talk as unintelligible. "What is the 'us' or 'me'? It is not the whole person from birth to death, the total space-time entity. Nor is it any particular temporal stage of the person," because at every temporal stage we have this same feeling, even though events regarded as "in the future" at one temporal stage of our lives are regarded as "in the past" at a later stage. As an event simply is what it is, it cannot be "in" *both* past and future. So these temporal categories seem to be meaningless.

Having argued that the flow of time is an illusion, Smart admits that "it is a strange and intellectually worrying one." What causes it? Is it a form of temporal dizziness connected with memory, or the flow of information through the brain? The problem with explaining this particular illusion is that it seems to be the illusion of an absurdity. It is not merely that our senses are misleading us, as in the case of a stick that looks bent in water. We seem to have a powerful impression of something, but a something that doesn't make any sense when scrutinized!

THE MYTH OF PASSAGE

They say more people are laid low by time anxiety than by time itself. But only time is fatal.

MARTIN AMIS

David Park is a physicist and philosopher at Williams College in Massachusetts with a lifelong interest in a time which he too thinks doesn't pass. For Park, the passage of time is not so much an illusion as a myth, "because it involves no deception of the senses. . . . One cannot perform any experiment to tell unambiguously whether time passes or not."[4] This is certainly a telling argument. After all, what reality can be attached to a phenomenon that can never be demonstrated experimentally? In fact, it is not even clear how to *think about* demonstrating the flow of time experimentally. As the apparatus, laboratory, experimenter, technicians, humanity generally and the universe as a whole are apparently caught up in the same inescapable flow, how can any bit of the universe be "stopped in time" in order to register the flow going on in the rest of it? It is analogous to claiming that the whole universe is moving through space at the same speed—or, to make the analogy closer, that *space* is moving through space. How can such a claim ever be tested?

Suppose you met an alien who claimed he had no idea what you meant by the flow of time. How would you describe it? What would you say to convince him of its reality? Ah, you may counter, an experience of temporal passage is an essential component of sentience! A being who had no notion of the passage of time would not be a true, conscious being like us at all. He/she/it could not even converse sensibly with us.

It is true that much of human concern turns on the passage of time: our hopes and fears, our nostalgia, our sense of destiny. From the great works of religion and literature down to the day-by-day organization of our lives, all human endeavor is a struggle enacted upon the river of time. Yet these are subjective and emotional aspects of life. When it comes to the truly objective properties of the world, reference to the flow of time appears superfluous. In fact, we *can* converse with the alien. True, we find it *convenient* to use time-flux language when discussing events in the objective world, but we do not need to do so.

Let me give an example of how we can purge our language of flux-speak. Meteorologists are among the most habitual users of flowing-time terminology. Typical examples are: "The summer will bring warmer weather," or "The rain will have stopped by the time Thursday comes." Perhaps this practice derives from the fact that weather systems flow across the globe as well as develop over time in fixed locations, so there is a tendency to slip unconsciously between references to movement in space and movement in time, or even to leave things deliberately ambiguous. "There is some stormy weather coming along" can mean either something like "Stormy weather is moving up from the south" or "Stormy weather is likely to develop over the city tomorrow."

Consider a flux-ridden statement like: "Only last Thursday, the weather bureau predicted that Saturday would be fine, but when the day came, the rain merely intensified, and it was not until Sunday morning arrived bright and sunny that I knew the worst was over." Although this informal description of a sequence of events colorfully conveys the necessary information, exactly the same essential content is contained, albeit more dryly, in the following catalogue of events: Thursday: the weather bureau issues a forecast predicting fine weather on Saturday. Friday: there is rain. Saturday: there is heavy rain. Sunday: there is sun. Notice I have used the present tense throughout, as a convenient way to correlate date labels with states of the weather. In fact, no verb is strictly necessary to state this correlation; we can imagine simply inspecting entries in a diary. The gist of the message that we get from this terse account is the same as the original version, but in the latter case nothing "happens" or changes, no days "come along" or "bring" good weather.

DOES THE ARROW OF TIME FLY?

Gather ye rosebuds while ye may
Old Time is still a-flying.

ROBERT HERRICK

Many people muddle the flow of time with the arrow of time. This is understandable, given the metaphor. Arrows, after all, fly—as time is supposed to. But arrows are also employed as static pointers, such as a compass to indicate north, or a weather vane to show the direction of the wind. It is in the latter sense that arrows are used in connection with time. In Chapter 9, I discussed the fumbling attempts by physicists to pin down the arrow of time. The quality this arrow describes is not the *flux* of time, but the asymmetry or lopsidedness of the physical world *in* time, the distinction between past and future directions of time.

Time does not have to *flow* from past to future for a time asymmetry to be manifested. To see why, imagine a movie film of a typical irreversible process: an egg falling on the floor and breaking. Suppose the film is chopped up into frames, and the frames are shuffled. Faced with the tedious task of reordering them, most people would have little difficulty in eventually restoring the original sequence, more or less. We suspect that frames of intact eggs come near the beginning and frames of broken shells near the end. The asymmetry of the sequence is obvious by inspection; it is not actually necessary to *run* the movie and see the ordered events "unfold" to discern the arrow of time. This arrow has nothing to do with the *movie;* it is a structural property of the set of frames. The arrow, or asymmetry, is just as much there if the frames are simply piled up on top of one another in sequence as it is if they are stuck together again and run through the projector.

I am as guilty as anyone for perpetuating the conflation of the flow of time and the arrow of time. In the foregoing chapters, I have unashamedly talked of time running faster in space, or time flowing backwards in another part of the universe. This was for stylistic convenience. "Time running faster in space" really means that the duration between two events as measured by a clock in space is a little greater than that measured by a clock on Earth. It is the interval of time between the events that is the issue, not some mythical temporal motion by which the world speeds from one event to the other. Likewise, "time flowing backwards" simply means the arrow of time reversing.

Of course, the existence of an arrow of time does not rule out a flow of time. Logically, however, if time did flow, it need not be in the direction indicated by the arrow. Time could flow from future to past, and an observer would then see events "going backwards" relative to our own experience of the world. On the other hand, if the flow of time is all in the mind, then it is likely that its direction will coincide with the arrow of time, because the arrow determines the directionality of thermodynamic processes in the brain. If so, then to say that time flows backwards when the arrow of time reverses is factually correct if by the statement one means that time *appears* to flow backwards.

The linguistic quagmire is exacerbated by the use of the words "past" and "future," which also have a dual meaning, as I mentioned in Chapter 9. Einstein demolished the absolute categories of *the* past, *the* present and *the* future, but past and future still retain some physical meaning in relativity theory. For example, it is still possible to say that one event occurred later than another, so event A might be in event B's future. This statement has nothing to do with event A or B actually "happening": the temporal relation between A and B is a timeless property, and unconnected with the existence of a now, or what moment of time a particular human being might decide "now" is in relation to A and B. As I stressed in Chapter 9, we can say that the arrow of time points (by convention) "towards the future," without implying that there is a *region* of time— "*the* future"—any more than when we say that a compass arrow points "towards the north" we mean that it is pointing at a particular place—*the* North. Instead, the arrow of time and the compass arrow indicate a *direction*—in time and in space, respectively.

WHY NOW?

Why do we not live in the reign of George III?

J. E. McTaggart

It is not merely temporal flux that baffles us. The passage of time is often viewed as the progress of "the now" *through* time. We can envisage the time dimension stretched out as a line of fate, and a particular instant— "now"—being singled out as a little glowing point. As "time goes on," so the light moves steadily up the time line towards the future. Needless to say, physicists can find nothing of this in the objective world: no little light, no privileged present, no migration up the time line.

So where has the now gone? As a child, I was deeply shocked when my mother told me that had she not met my father I would not have been born. It had never occurred to me. Of course, she may still have given birth to a baby in 1946, but that baby would not have been *me*—it would have been someone else! What then? My childhood sense of "she'll be right" led me to suppose that I would have been born at some other time, to some other parents. But when? I lay awake at night pondering this. Why was I living *now* rather than at some other epoch in history? I could so easily have found myself living in Roman times, or in the twenty-fifth century. Given that I *must* exist, what, I wondered in bewilderment, determines *when* I exist? For me, "now" is when I am alive to experience the world. So why is it the twentieth-century *now?* In other words, why is it "now" *now?* Is there something special about *this* now—*my* now—as opposed to other nows, like twenty-fifth-century ones? Will twenty-fifth-century worriers puzzle over what is special about *their* now too?

Unless, of course, there isn't anybody left by then.

Ah! Could this explain why I am alive *now* . . . because I couldn't be alive *then?* Or, turning the argument on its head, could the fact that I *am* alive now imply something unpleasant about the human species in the twenty-fifth century?

Brandon Carter, a British astrophysicist who lives in France, had something to say on the matter. Among fellow astrophysicists, Carter is famed for his work on black holes. Others know his name in connection with something called the "anthropic principle." This states, innocuously enough, that the world we see around us cannot be such that it forbids conscious beings. Seeing as we *are* here, and conscious, it is no surprise that we observe a world consistent with our existence. It could hardly be otherwise. In this form, the anthropic principle is trivial. But it gets more interesting when we take into account that some aspects of what we observe may not be typical of the whole. For example, our location in space is hardly typical. Most of the universe is either a vacuum or a tenuous gas, yet we live on the surface of a solid planet. Most planets are very hot or cold, yet ours is equable. There is nothing mysterious about this: the existence of conscious biological organisms requires special circumstances, like a solid planet at a suitable temperature. We could not have evolved anywhere very different. It may also be that our sun, or the Milky Way galaxy, is in some way special (there is no observational evidence that they are). If so, this would provide a reason for why we find ourselves living in *this* part of the universe rather than some other.

It is but a small step from recognizing that our location in space is untypical to reaching the same conclusion about our location in time. Perhaps we are living at this epoch rather than some other because life would be impossible at other epochs? The American astrophysicist Robert Dicke many years ago pointed out that life (at least of the sort we know) requires certain key elements, such as carbon, and it is unlikely that these existed right after the big bang.[5] Carbon was not there in the beginning, but is manufactured inside stars. The stars can disgorge their carbon into space in a number of ways, most obviously by supernova explosions, so the carbon gets continually recycled into new generations of stars and planets. Dicke reasoned that it would take at least the time needed for one generation of stars to live and die before biological life could get started. On the other hand, after a few stellar generations had elapsed, stars would start getting scarce, and equable planetary systems like the solar system would be a thing of the past. It follows that our existence at this epoch (about two or three stellar generations into the great cosmic drama) is fairly typical—and no surprise.

At a memorable meeting at the Royal Society in London in 1983, Brandon Carter advanced this "why now?" theme a dramatic (and, in the opinion of many, absurd) step further. Imagine all the human beings who will ever have lived, he said. If humanity survives its present troubles and thrives for thousands or even millions of years, nearly all the people who ever live will live a long time in our future. So *we* would be very untypical humans, living as we do at the end of the twentieth century. But what reason have we to suppose that we late-twentieth-century folk—mere random human beings in the vastness of human history—are *special?* None. Hence: the assumption that humanity will thrive for a long duration is suspect. If we *are* typical, then humanity is doomed, and destined for imminent annihilation.

Perhaps because this apocalyptic prediction was delivered in a seriously downbeat tone and obscured, as I recall, with almost unreadable slides, it fell on largely deaf ears at the time. Carter himself didn't push the argument, but felt that nuclear-submarine commanders would do well to reflect upon it.

Carter's bold confrontation of the why-am-I-living-now teaser rekindled my childhood perplexity. Look at the three graphs shown in Fig. 12.1. These show three possible scenarios for the future of mankind, based on alternative projections of population growth. In (a) the number of humans goes on rising into the far future. It is hard to see how this can come about without rapid colonization of other planets. In (b) the Earth's population rises strongly, then stabilizes, perhaps at twenty billion. In (c) the number peaks at a value not a lot higher than the present value, and then falls sharply. Each graph shows our own approximate position ("now"), which coincides with the sharp population rise characteristic of the twentieth century. It is interesting to note that, because of this accelerating population growth, about 10 percent of people who have ever lived are living now.

It is obvious at a glance that to be living right on this steep slope is highly untypical in both scenarios (a) and (b), but pretty typical in the case of (c). This suggests (c) may be close to the true distribution of humans, and predicts that the peak probably isn't too far in the future. The subsequent drastic population decline could come about in many ways—nuclear war, disease, eco-disaster, asteroid impact, etc., etc.

Most people shrug Carter's argument aside with scorn. How can we possibly predict the future of free human beings from imaginary graphs and arguments about probability? Those future beings don't even exist yet. How can we place their observations (e.g., twenty-fifth-century nows)—or perhaps their nonobservations—on the same footing as *our* observations *now* (i.e., at *this* now)? After all, we *really exist* now; they don't yet exist, do they?

Anyone who has read Chapter 2 carefully will know that this is a flaky objection. Einstein scuttled the idea of a universal now, and pointed the way to "block time," in which all events—past, present and future—are equally real. To the physicist, human beings of the twenty-fifth century *are* "there" (or not, if curve c in Fig. 12.1 is a correct prediction). They are there—*in the future!*

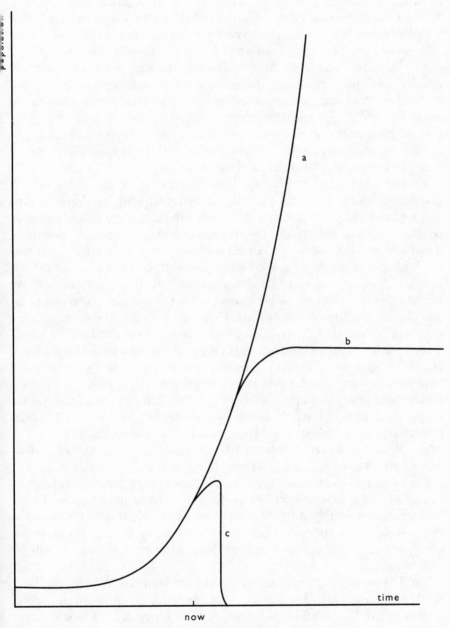

12.1 Doom soon? The graphs show three alternative projections for human population growth. The common feature is the sharp rise in population in the twentieth century. Our existence at this time is atypical and coincidental except in the case of curve c, where the escalating population is soon to be chopped back, perhaps by a sudden catastrophe.

Though Carter was coy about his Doomsday argument (he omitted it from the published version of his lecture[6]), the Canadian philosopher John Leslie has written extensively about it.[7] Leslie likens the collection of all humans who will ever have lived to counters resting in a vast imaginary urn. There is a counter there for each of us, inscribed with our name. The great hand of fate dips into the urn and draws counters one by one, thus conjuring the corresponding person into being. We know that about forty billion counters have been drawn so far (there are about four billion people alive today). Based on the evidence available, can we conclude anything about the number of counters that remain in the urn? Carter and Leslie say yes, and argue that it is unlikely to be a much larger number than the total drawn so far.

To see why, let's cut the numbers down to everyday urn size and consider a simple practical experiment. Suppose you are shown an urn and told that there are two possibilities: (i) the urn contains ten counters (Carter's doomy view), and (ii) the urn contains a thousand counters. You have no idea which of (i) or (ii) is the actual state of affairs, but you are told that in either case your name is inscribed on one and only one counter. You are asked to bet on (i) versus (ii). In the absence of any evidence at all, it's anybody's guess. Let's say you are skeptical of (i) and decide it's fifty to one against. Counters are now drawn out one by one, and by the time three have been drawn your name has appeared. You are asked if you would like to revise your estimate of the odds. Well, of course you would! You are gambling heavily on the urn's containing a thousand counters, and yours has been drawn after only three. That is much more likely to happen if there are only ten counters in all rather than a thousand. In the light of this new evidence, there is a well-known mathematical formula called "Bayes's rule" for computing the best-bet odds. With the above numbers it turns out that the probability of (i) being correct is now $2/3$, i.e., twice as likely as (ii).

Bayes's rule is a standard technique for assigning probabilities to competing hypotheses when we are given only limited information. Leslie believes we can apply it to the case of the Great Human Urn and, using the evidence that "our counters" have been drawn fairly early on in the piece, we can reasonably conclude that Carter is right, and it will be Doom Soon.

The Doomsday argument is bolstered from an unusual direction. Isn't it odd, asked Carter, that "now" happens to be about when the sun is middle-aged? If evolution had gone just a bit slower, we would never have made it in time. The sun would have burned steadily for a few billion years; life would have arisen on Earth and progressed a bit, and then been snuffed out by the death throes of our star before any sentient being emerged to worry about it. Because the processes of biological evolution are largely accidental and have no obvious connection with the processes that determine how fast the sun ages, there would seem to be no physical link between the lifetime of the sun and the time scale of evolution. The fact that these long durations nevertheless turn out to differ only by a factor of about two seems highly suspicious.

Carter explains the "coincidence" that these two seemingly independent time scales roughly concur, using a curious argument. It must be, he reasons, that intelligent beings like us are exceedingly improbable— so improbable in fact that you would expect to wait a truly vast length of time for them to evolve. This doesn't mean they *can't* arise sooner (obviously, because they have)—a very rare random process can always occur more promptly by chance, against the odds—but it is likely that those odds will be beaten later rather than sooner, to allow the maximum duration of time for whatever rare sequence of accidents is needed to occur.

To make this point clear, let me consider another gambling analogy. Suppose you are asked to throw three dice and obtain three ones, followed at some stage by three twos, followed at some later stage by three threes. These improbable triples correspond to some unlikely step in the evolution of mankind (e.g., the emergence of consciousness). You are allowed fifty throws in all. The chances are, you won't throw the required sequence at all, but if you *do*, it is more likely that the three-threes step will occur near the end of the trial than at the beginning, to allow the maximum number of goes for the earlier two steps to be completed first, against the odds. To see why, note that the chances of obtaining three ones, twos and threes in the *first* three throws are much less than if the whole fifty throws are used up trying.

Reworking these dicey ideas in the context of human evolution, if there are n unlikely steps in our development, the bigger the number n, the closer we will probably be to the "end of the trial"—i.e., Doomsday. Now, some biologists might argue that n is only one or two. If that is correct, then the curious fact that the epoch of human existence turns out to be (to within a factor of two) the same as the sun's total life expectancy is comfortably explained. On the other hand, you could forget about the sun and the time-scale coincidence and turn the argument on its head, reasoning that n is very large. Most biologists, in fact, think n is a big number—that there were a great many improbable accidents that went to make up *Homo sapiens*. If they are right, then we are probably very close to Doomsday. Carter was able to provide a handy formula, based on simple probability theory, for how long we can expect to survive. To compute the time left, you divide the expected total solar lifetime (say, eight billion years) by $n + 1$. If n is one million, we can expect no more than about eight thousand years before being annihilated by one means or another. (For a list of horrible alternatives, see my book *The Last Three Minutes*.)

If you believe Carter's argument, then you don't need to worry why you are a human living now rather than a little green person in the Andromeda galaxy in a hundred billion years' time. The chances are that there won't be any such persons, green or otherwise. Though the possibility of some lowly form of life elsewhere is left open, the argument suggests that intelligent life is pretty well confined to Earth at this epoch —which provides a unique and improbable window of opportunity in space and time which we just happen to occupy by a fluke.

And *now* for something completely different . . .

CHAPTER 13
EXPERIMENTING WITH TIME

Time travels in divers places with divers persons. I'll tell you
who Time ambles withal, who Time trots withal, who Time
gallops withal, and who he stands still withal.

WILLIAM SHAKESPEARE

HOW LONG DOES THE PRESENT LAST?

*We have no reason to think that even now time is quite perfectly
continuous and uniform in its flow.*

C. S. PEIRCE (1890)

Until now, I've been writing about "now" as if it were literally an instant
of time, but of course human faculties are not infinitely precise. It is
simplistic to suppose that physical events and mental events march along
exactly in step, with the stream of "actual moments" in the outside
world and the stream of conscious awareness of them perfectly synchronized. The cinema industry depends on the phenomenon that what seems
to us a movie is really a succession of still pictures, running at twenty-
five frames per second. We don't notice the joins. Evidently the "now"
of our conscious awareness stretches over at least $1/25$ of a second.

In fact, psychologists are convinced it can last a lot longer than that.
Take the familiar "tick-tock" of the clock. Well, the clock doesn't go
"tick-tock" at all; it goes "tick-tick," every tick producing the same
sound. It's just that our consciousness runs two successive ticks into a
single "tick-tock" experience—but only if the duration between ticks is

less than about three seconds. A really big pendulum clock just goes "tock . . . tock . . . tock," whereas a bedside clock chatters away: "ticktockticktock . . ." Two to three seconds seems to be the duration over which our minds integrate incoming sense data into a unitary experience, a fact reflected in the structure of human music and poetry. In his essay "The Dimension of the Present Moment," Hungarian poet Miroslav Holub reports that, in 73 percent of all German poems read aloud, the lines last from two to three seconds—the basic "sound bytes" are deliberately adapted to the speed of our mental functions. Poems with longer lines are read with a slight unconscious pause in the middle of each line. I have no doubt the same would be found for English poetry. "In this sense," writes Holub, "our ego lasts three seconds. Everything else is either hope or an embarrassing incident."[1]

On the other hand, human beings can certainly perform some conscious tasks, like slamming on the brakes of a car, on a much shorter time scale than this. In an activity like playing the piano, the finger movements are performed at film-blurring speed in response to an overall concept of "the tune": the pianist is not conscious of instructing each movement individually. Perhaps there are many "nows" of varying duration, depending on just what it is we are doing. We must face up to the fact that, at least in the case of humans, the subject experiencing subjective time is not a perfect, structureless observer, but a complex, multilayered, multifaceted psyche. Different levels of our consciousness may experience time in quite different ways. This is evidently the case in terms of response time. You have probably had the slightly unnerving experience of jumping at the sound of a telephone a moment or two before you actually hear it ring. The shrill noise induces a reflex response through the nervous system much faster than the time it takes to create the conscious experience of the sound.

It is fashionable to attribute certain qualities, such as speech ability, to the left side of the brain, whereas others, such as musical appreciation, belong to processes occurring on the right side. But why should both hemispheres experience a common time? And why should the subconscious use the same mental clock as the conscious? It is sometimes claimed that dreams "run" at very high speed relative to corresponding waking experiences, though I know of no convincing experimental evidence for this. However, certain mental states are definitely associated with altered rates of passage. The neurologist Oliver Sacks once described to me an experience in which he swam in preoccupied mood for many hours, believing only an hour or so had elapsed, and emerged from the water exhausted. Sensory deprivation can also drastically alter the impression of time intervals. Practitioners of meditation claim that they can more or less suspend their perception of the flow of time altogether by detaching themselves from worldly events.

Psychologists have devised some ingenious ways to help unpack the human "now." Consider how we run those jerky movie frames together into a smooth and continuous conscious stream. This is known as the "phi phenomenon." The essence of phi shows up in experiments in a darkened room where two small spots are briefly lit in quick succession, at slightly separated locations.[2] What the subjects report seeing is not a succession of spots, but a *single* spot moving continuously back and forth. Typically, the spots are illuminated for 150 milliseconds separated by an interval of fifty milliseconds. Evidently the brain somehow "fills in" the fifty-millisecond gap. Presumably this "hallucination" or embellishment occurs after the event, because until the second light flashes the subject cannot know the light is "supposed" to move. This hints that the human now is not simultaneous with the visual stimulus, but a bit delayed, allowing time for the brain to reconstruct a plausible fiction of what has happened a few milliseconds before.

In a fascinating refinement of the experiment, the first spot is colored red, the second green. This clearly presents the brain with a problem. How will it join together the two discontinuous experiences—red spot, green spot—smoothly? By blending the colors seamlessly into one another? By waiting until the visual stimulus of green arrives before flipping? Or something else? In fact, subjects report seeing the spot change color abruptly in the middle of the imagined trajectory, and are even able to indicate exactly where using a pointer. This result leaves us wondering how the subject can apparently experience the "correct" color sensation *before* the green spot lights up. Is it a type of precognition? Commenting on this eerie phenomenon, the philosopher Nelson Goodman wrote suggestively: "The intervening motion is produced retrospectively, built only after the second flash occurs and projected backwards in time."[3] In his book *Consciousness Explained*, philosopher Daniel Dennett points out that the illusion of color switch cannot actually be created by the brain until after the green spot appears. "But if the second spot is already 'in conscious experience,' wouldn't it be too late to interpose the illusory content between the conscious experience of the red spot and the conscious experience of the green spot?"[4]

NOW YOU SEE IT, NOW YOU DON'T

The word Time came not from heaven but from the mouth of man.

JOHN WHEELER

Common sense suggests that in the two-spot experiment the brain waits until it sees the green spot, then "goes back" and fills in the transition, finally delivering up the smoothed-out, edited package to "the conscious subject" as a consistent whole, a short while after it all "actually happened." Dennett refers to this explanation as the "Stalinesque model," comparing it to a censor in a radio control room who installs a loop in the recording tape to delay transmission for a few seconds in order to bleep out obscenities. The trouble is, it takes two hundred milliseconds from the beginning of the experiment until the appearance of the green light, and this is long enough to notice a "gap in the film" (it corresponds to about five frames of a real movie). Certainly subjects can respond much faster than this to visual stimuli if they are primed to do so. For example, it is possible for the subject to press a button signaling that he or she has become aware of the presence of the red spot well before the green spot appears. So it is hard to see how consciousness can be made to "mark time" for that long.

Dennett considers an alternative explanation—the Orwellian one. Here the subject initially has the conscious experience of the red spot, then of the green spot, but a sort of inbuilt censor, putting one and one together and failing to obtain two, proceeds to edit the original account of events, replacing it by one containing a smooth trajectory. This entails wiping out the offensive original memory with the break in the illumination, and replacing it by the edited version of a continuous trajectory.

If the Orwellian explanation seems too fantastic, consider another series of experiments discussed by Dennett. In this case, an image (a disc) is flashed before a subject—briefly, but for long enough to be perceived and correctly reported. If, shortly afterwards, a second image (a ring, positioned to envelop the spot that the disc previously occupied) is flashed, it has the effect of erasing the perception (or at least the memory of the perception) of the first image, almost as if the brain, on receiving the second image, decided to censor the first.

Both the Stalinesque and Orwellian explanations are based on the common notion that there exists a "seat of consciousness"—a subject lurking somewhere in our brains like a spectator in a movie theater— being fed an edited stream of sensory impressions, complete with temporal tinkering. On this view, which dates back to René Descartes's dualistic theory of mind/brain, you see something "happen" when the brain (after due processing of the data) "presents the finished product" to "you"—the spectator. On this view a definite line can be drawn in time: the moment the sense data "enter your consciousness." Unfortunately, such a picture of consciousness is almost totally discredited these days. Dennett, for example, prefers to think of consciousness in terms of what he calls a Multiple Drafts model, involving a series of processing and editing functions that continually refine and update incoming data in parallel, smearing the temporal relations as a result. In this theory, there is no movie theater, no show, no spectator, and no seat of consciousness where "you" become aware, at some instant of time, of some real event that "just happened" in the outside world. Instead, both "you," and your awareness of a stream of events in a certain time sequence, are *created* by the confluence of these many parallel streams of data processing.

Dennett denounces the Stalinesque notion that certain bits of data course around the brain beyond our ken, possibly getting massaged on the way, and then finally enter some sort of charmed circle where we become conscious of them. He is equally scathing about the Orwellian view, where data sometimes enter our consciousness only to be summarily snatched out again, causing the information to vanish no sooner than it had arrived. Conscious awareness, claims Dennett, is not the presentation of (possibly edited) data to a mythical subject (mind), but *is* the sum total of the data streams taken together. "The brain doesn't actually have to go to the trouble of 'filling in' anything with 'construction,' " he writes, "for no one is looking." [5]

If the brain is not, after all, a type of projector "putting on a show" for a spectator (you, the conscious subject), then the problem of projecting "backwards in time" as Goodman expressed it, simply evaporates away. In the Multiple Drafts theory, there is nothing so clear-cut as a one-to-one matching of physical events with mental events. The subject builds up a narrative about the world from a range of information streams (drafts) which are continually subject to editing and even retraction. Each stream may provide its own time line, to be placed alongside the time line of objective events. It may often happen that an informational time line "loops back" for a few milliseconds relative to other informational time lines, or to "objective time," thereby further refining the editing process. The result is the compelling illusion of a smooth and consistent meta-narrative being presented to an independent spectator.

FILLING IN TIME

There exists, therefore, for the individual, an I-time, or subjective time. This in itself is not measurable.

ALBERT EINSTEIN

Dennett also discusses an experiment in which the brain seems to play temporal tricks with tactile information. The subject wears a device that delivers light taps to the arm in a certain sequence: a few to the wrist, followed by a couple to the elbow, then the shoulder, in rapid succession. When this is done, the subjects report the sensation of equispaced taps traveling up the arm, like a little animal hopping. In other words, some taps are felt to occur *in between* the points of contact, such as on the forearm. Once again we have the mystery of how the brain knows it is going to receive an elbow tap after the wrist tap, in order that it may create the false impression of a forearm tap in between. Is this a case of backward causation? Not so, says Dennett, merely another example of parallel processing, where different narratives of the event sequence stream through the brain, getting edited, compared and rejected, and eventually blending to create the illusion.

That there must be *some* integration of parallel data streams going on in the brain all the time is obvious from the fact that nerve impulses from different parts of the body arrive at the brain at different moments. These impulses travel relatively slowly, but survival may depend on our ability to respond rapidly. In such tasks as hand-eye coordination, the brain cannot afford the luxury of parking some impulses in a sort of holding bay while it awaits the arrival of others, to ensure synchrony. Under this time pressure, the brain must continually be "ahead of events," building a picture of a sometimes surprising world based on fragmented information that is subject to continual revision. There may be a sound biological *need* for a reversal of the order in which sense data arrive and the time order that the subject infers.

Oft-cited in this connection are the neurosurgical experiments conducted by Benjamin Libet of the University of California.[6] During brain operations, the patient is usually kept conscious. Libet took the opportunity to attach electrodes to exposed brains. By stimulating the cortex electrically, he was able to produce the sensation of a tingle in the patient's hand. In the experiment, Libet attached electrodes to the skin of the hands too, so he was able to compare the experiences of tingles reported by the patient when the hands and the cortex were both stimulated.

In the first part of the experiment, Libet found that the actual sensation of a tingle occurred up to half a second after the stimulus was delivered to either the hand or the cortex, even though the signal's travel time to the brain was only about ten milliseconds. The results of the second part of the experiment were even more surprising. Libet tried stimulating the left hand at the same moment as the left cortex. The latter produced a tingle in the *right* hand, so the patients felt tingles in both hands and were able to report on which one seemed to occur first. Now, you might suppose that the cortex was somehow closer to the "seat of consciousness" than the hand, so the brain-induced tingle in the right hand would be experienced before the skin-induced tingle in the left hand. But the time order was completely reversed! The patients definitely felt the left hand tingle first. Even when the hand was stimulated a short while *after* the brain, the order was reversed.

Libet explained his unexpected results by claiming that, when the skin is stimulated, the sensation experienced half a second or so later is "referred back in time" to when it actually occurred, whereas no such backward referral takes place for cortical stimulation. This makes it *seem* as if the skin were stimulated first, whereas in fact it is the other way around. A naïve interpretation of the experiment is that we become conscious of at least some events in the world about half a second after they actually happen—i.e., the personal "now" is half a second "late," and the perceived world is really a sort of action replay.

Libet's work suggests that a substantial delay occurs when the subject behaves as a passive observer. A similar delay seems to be involved when brain activity occurs in the other direction—that is, when a person freely wills a voluntary action. Electrodes attached to the scalp can monitor brain "waves" and detect bursts of activity that are associated with voluntary movements, such as flexing a finger. A German research team directed by H. H. Kornhuber found that in some cases brain cells start firing as long as a second or more before the physical movement actually begins.[7] It is almost as if your brain knows what "you" are going to do some moments before you decide to do it! Or, at least, the brain "gets going" on the task before you *think* you decided to do it. This precursory electrical burst has been dubbed the "readiness potential" by the philosopher Karl Popper and the neurophysiologist Sir John Eccles, who argue, in a throwback to Descartes's dualism, that it is caused by a nonmaterial mind somehow priming the brain to do its bidding.[8]

In considering the Libet and Kornhuber experiments, Roger Penrose points out that, taken together and at face value, they imply something quite startling.[9] Given that consciousness runs over half a second late and actions require the brain to "get ready" for up to a second or more, it would seem as if a human being cannot consciously respond to an event in less than about a couple of seconds. This certainly contradicts experience. People can respond in fractions of a second to a stimulus if asked to do so. But this might mean you are acting as a mere automaton when you make such rapid responses, and only *imagine* you are using conscious will. Or maybe our concepts of time and human will, being so obviously interwoven, are much more subtle than the above simplistic picture suggests.

Consciousness of time differs from consciousness of other physical qualities, such as spatial size or shape, in a significant respect. When we see a shape such as a square, the electrical activity in our brains is not square-shaped. There is no "little square" inside our heads, projected on a movie screen for the subject to watch. Instead, a complicated pattern of electrical activity (somehow!) produces the sensation "square." That is to say, the square is *represented* by an electrical pattern. We must not confuse the pattern of the representation with the shape being perceived; the representation is *not* the same pattern as the object. When it comes to time, however, the situation is more complicated. Our first thought is that a time sequence of events in the external world is represented in our brains by a matching time sequence of electrical pulses. This would be the temporal equivalent of the "little square," with the pattern of electrical activity tracking the succession of events "out there." But the fact that the brain receives a desynchronized jumble of signals from which it has to build up a consistent impression of time suggests otherwise. It may be that the electrical patterns in the brain that *represent* time sequences may be quite different from the *actual* time sequence of the events they represent.[10]

However, there is *something* special about time in this respect. First, there *are* cases where the electrical sequence in the brain would seem to shadow the time sequence of events in the outside world. For example, the motion of an object from left to right through the visual field, even for a very brief duration, can be distinguished from the reverse motion. Second, the conscious subject is not just a passive observer but an agent. The incoming signals serve not merely to inform us, but to goad us into action: they have causative effects. Nerve signals are physical things, and as such are subject to the laws of physics, like everything else. The time order of *physical* events does matter: we cannot act on sensory information until we have it. So the "sorting out" of temporal sequences had better not take too long, or we could be hit by a falling branch while we mull the situation over. It follows that the human "present moment," revealed now in all its psychological and physiological complexity, cannot last for more than a fraction of a second.

SUBJECTIVE TIME

Bees are not as busy as we think they are. They just can't buzz any slower.

KIN HUBBARD

Whatever the explanation for the fascinating experiments I have mentioned (and I have no doubt it is not a straightforward one), most psychologists agree that framing a clear concept of time is a higher mental function. It is possible that humans are alone in having such a richly developed notion of temporality. Of course, some of the baser aspects of temporal experience must be common to many animals, and derive from the various internal biological clocks that regulate organic activities. The biologist Stephen Jay Gould has drawn attention to the interesting fact that these clocks, and hence the *pace* of life, depend sensitively on bodily size:

We are trained from earliest memory to regard absolute Newtonian time as the single valid measuring stick in a rational and objective world. We impose our kitchen clock, ticking equably, upon all things. We marvel at the quickness of a mouse, express boredom at the torpor of a hippopotamus. Yet each is living at the appropriate pace of its own biological clock.

Small mammals tick fast, burn rapidly, and live for a short time; large mammals live long at a stately pace. Measured by their own internal clocks, mammals of different sizes tend to live for the same amount of time.[11]

The pace of activity referred to here includes the rates of breathing, the heartbeat and the metabolization of food. These functions show precise mathematical scaling laws with body weight, as does life expectancy. Thus your pet mouse's heart beats several times faster than your own, but it is likely to be dead in a couple of years. The interesting question is, does the mouse's allotted two years *feel* to it like our three score and ten years does to us? In other words, does psychological time "run" at different speeds for mice and men?

I argued in Chapter 8 that the answer must depend more on the speed of thought than on physical reflexes or muscular functions. As far as I know, all mammals have roughly the same "speed of thought" (as measured by rate of neural activity), so it seems the poor old mouse really does have a short, if hectic, life. The same may not be true of an intelligent computer like Tipler's superbrain, which could operate much faster than a human, or an alien whose metabolic and neural processes may be completely different. If the alien's subjective experience of the passage of time depends on the rate at which information is processed, as Tipler and Dyson assumed (see Chapter 8), then, the faster the processing, the more thoughts and perceptions the alien will experience per unit time— and the faster time will appear to pass. This assumption is employed in an entertaining manner in the science-fiction novel *Dragon's Egg* by Robert Foreword, which tells the story of a community of conscious beings who live on the surface of a neutron star.[12] These compact aliens utilize nuclear rather than chemical processes to sustain their existence. Because nuclear reactions occur much faster than chemical reactions, the neutronic beings process information very swiftly. One minute on the human time scale represents the equivalent of many years for the aliens. In the story, the neutron-star community is fairly primitive when first contacted by humans, but develops before their very eyes and rapidly overtakes humanity.

Attractive though this simple view of psychological time may be, it is undoubtedly a gross simplification. Subjective impressions of time are clearly more than just a measure of the rate of brain activity, as the experiments of Philadelphia psychologist Stuart Albert prove. He shut volunteers in a room in which the wall clock had craftily been adjusted to run either at twice the speed or half the speed, without informing the subjects.[13] Amazingly, they were completely unaware of the deception; their mental functions adapted automatically to the accelerated or retarded pace. For example, memory was tested, and found to decay faster for subjects in the speeded-up group than the slowed-down group. Estimates of durations were similarly corrected, being scaled down for the "fast" people and up for the "slow" ones. Though our basic mental and physiological functions are regulated by reasonably accurate neurological and chemical clocks buried within us, it would seem that these clocks do not relate too closely to temporal *awareness* as such.

I believe our conscious experience of time is more likely to be associated with a sense of personal identity—a concept that evolved much later than the more primitive biological and cognitive cycles, along with language, art and culture. It is, then, all the more amazing that this complex and sophisticated notion—time—should find itself playing so profound a role in the objective description of the physical universe. Mathematics and time are the two great abstractions that kick-started science as we know it. Both are products of the higher human intellect. How astonishing that these highly derivative concepts should find so fruitful an application to the fundamental processes of nature. Galileo, Newton and Einstein all chose time as the central conceptual pillar around which to build a scientific picture of physical reality, and yet, when we stare into our own minds to find the foundation of temporal experience, it seems to crumble away, leaving only mystery and paradox.

The enigma of time's stunning scientific utility has been eloquently phrased by the Japanese philosopher Masanao Toda:

> No one, apparently, can claim to know what time is. Nevertheless, there is this brave breed of people called physicists, who used this elusive notion as one of the basic building blocks of their theory, and miraculously, the theory worked. When one of the leading figures of the clan, by the name of Albert Einstein, quietly mumbled his secret incantation which sounded like "Combine time with space in such a way that nothing can travel faster than the speed of light, then mass is equal to energy," lo and behold, atoms exploded ever so noisily.

Einstein's time is clearly part of the truth. But is it the whole truth? Toda, for one, thinks not:

> There is no doubt that physicists succeeded in trapping some really important ingredient of time within their capsule labeled *t*, but equally certain it is not all of the time that is captured within their capsule. Our intuition is crying out to tell us that time is something that flows unlike the physical time which is frozen still.[14]

THE BACK DOOR TO OUR MINDS

Clock time is our bank manager, tax collector, police inspector; this inner time is our wife.

J. B. Priestley

Time is the mediator between the possible and the actual.

G. J. Whitrow

In appropriating time for themselves, and abstracting it into a stark mathematical parameter, physicists have robbed it of much of its original, human, content. The physicist will usually say, "Ours is the *real* time—and all that there really is. The richness of human psychological time derives entirely from subjective factors and is unrelated to the intrinsic qualities of real, physical time"—and then go about his or her work and daily life immersed in the complexities of human time like everyone else.

Should we simply shrug the human experience of time aside as a matter solely for psychologists? Does the time of an altered state of consciousness have no relevance at all to the time of Newton or Einstein? Does our impression of the flow of time, or the division of time into past, present and future, tell us nothing at all about how time *is* as opposed to how it merely appears to us muddle-headed humans?

As a physicist, I am well aware how much intuition can lead us astray. As I remarked earlier, intuition suggests that the sun moves around the Earth. Yet, as a human being, I find it impossible to relinquish the sensation of a flowing time and a moving present moment. It is something so basic to my experience of the world that I am repelled by the claim that it is only an illusion or misperception. It seems to me there is an aspect of time of great significance that we have so far overlooked in our description of the physical universe.

In this unease I am clearly not alone. Many scientists have suggested that there should be some subtle physical process that "makes time flow," or at least appear to flow. Scientists are divided as to whether the process concerned is a general one that bestows a flow of time on the universe as a whole, or is merely something odd restricted to the human brain which gives us a sense of the passage of time. Prigogine, for example, belongs to the first camp, and has suggested that the traditional laws of motion of material particles, which are time-reversible, are wrong and should be replaced by slight modifications which build in a temporal directionality at the most basic level.[15] In the second camp are physicists

such as Penrose who maintain that the answer lies in quantum physics
and the still-mysterious processes in the brain that accompany acts of
observation of the world.[16]

This anxious search for a "missing link" between flowing subjective
time and the frozen block time of the physicist has a long history. We
have seen how the Greek philosophers emphasized the distinction between *being*—the quality of enduring existence—and *becoming*—the
quality of change or flux in physical systems. In the 1920s, Eddington
declared that our impression of becoming, of a flowing time, is so powerful and central to our experience that it must correspond to *something* in
the objective world: "If I grasp the notion of existence because I myself
exist, I grasp the notion of becoming because I myself become. It is the
innermost Ego of all which *is* and *becomes*."[17] Eddington suggested that
we experience time in two distinct ways. The first is through our senses,
in the same manner that we perceive spatial relationships. But there is a
second way too, a sort of secret "back door" into our minds that enables
us to feel time directly, deep within our souls:

> When I close my eyes and retreat into my inner mind, I feel myself
> *enduring*, I do not feel myself *extensive*. It is this feeling of time as affecting
> ourselves and not merely as existing in the relations of external events
> which is so peculiarly characteristic of it; space on the other hand is always
> appreciated as something external.[18]

More recently, Roger Penrose has also written about "inner time" in
essentially the same way as Eddington:

> It seems to me that there are severe discrepancies between what we
> consciously feel, concerning the flow of time, and what our (marvellously
> accurate) theories assert about the reality of the physical world. These
> discrepancies must surely be telling us something deep about the physics
> that presumably must actually underlie our conscious perceptions. . . .[19]

Thus the flow of time, so basic to our experience, hangs as a tantalizing
mystery. Some, like Jack Smart, would have us sweep it under the
carpet, defining it away as a misuse of speech or simply dismissing it as
an illusion. Although I think that Smart has greatly clarified the subject,
at the end of the day I am forced to agree with Eddington and Penrose
that we are missing something important from the physics of time and
our perception thereof. There is no obvious "time organ" in our bodies
in the same sense as we possess "sight organs" and "sound organs."
Yet there *is* an inner sense of time—a back door—buried deep within
human consciousness, intimately associated with our sense of personal

identity and our unshakable conviction that the future is still "open," capable of being molded by our chosen actions.

It is ironical that Einstein's time, having placed the observer in a central role, makes no provision for the personal experience of flux, or the sense of past, present and future. In this respect it differs little from the time of Newton and Laplace. Like Laplace, Einstein was at heart a determinist. He found quantum physics, with its inherent uncertainty and indeterminism, utterly repugnant. Yet, as I pointed out in Chapter 1, a deterministic world is one in which the future is already contained in the present and nothing genuinely new ever happens. In such a world the division of time into past, present and future is a meaningless exercise, because the state of the universe at one moment contains all the information about its states at later moments. The "unfolding" of the future is nothing more than the outworking of pure logic through the mathematical laws of dynamics. As Laplace himself remarked in 1819, a superintelligence with complete knowledge of a deterministic universe would have no sense of the flow of time: "the future and the past would be equally present to its eyes."[24] Einstein's time, despite its limited observer-dependence, still adheres to Laplace's determinism, to a rigid chain-mesh of cause and effect, in which the destiny of the world has been etched into the fabric of nature since the dawn of existence.

If we identify Einstein's theory of relativity with the modern era of physics, then I contend that modern physics will not solve the riddle of time. But *postmodern* physics might. Two areas of investigation look promising. One is chaos theory, the other quantum mechanics. Both introduce a type of indeterminism into nature. A chaotic system is one which, although in a strict mathematical sense deterministic, is nevertheless so highly sensitized to minute disturbances that meaningful prediction over the long term is precluded. The tiniest disturbances amplify and amplify until they wreck the predictability of the system; its behavior is essentially random. Chaos theory suggests that many physical systems are chaotic, but some, like the human brain, operate at "the edge of chaos," a fascinating and ill-understood regime that combines novelty and openness with orderly operation, allowing the system to explore a rich repertoire of alternative states without descending into anarchy. This seems to capture some elements of human free will.

Quantum physics, like relativity, also places the observer in a central role, but in an altogether more significant fashion. The act of observation in quantum physics serves to concretize an otherwise fuzzy and uncertain physical state. As I have explained, quantum states generally involve multiple overlapping phantom realities. More accurately, these alternative worlds are *contenders* for reality—statistical expectations rather than actually existing physical universes—melded into a subtle amalgam. In the absence of observation, this cocktail of superimposed worlds evolves as a whole, but when we *inspect* events in the quantum domain, we see a specific, concrete, single reality, not a ghostly superposition of worlds. This "collapse" of multiple possibilities, of statistical expectations, into a unique actuality remains one of the great unsolved puzzles of physics.

Many scientists are adamant that the "concretization" of quantum reality has nothing whatever to do with the mind, but others maintain that the mystery of the "collapse" and the mystery of consciousness are intimately bound up with each other. Eddington and Bondi, for example, and philosophers such as Hans Reichenbach and Gerald Whitrow, have argued that the flow of time, or the phenomenon of "becoming," has its roots in this quantum "collapse" process. Thus, according to Hermann Bondi:

> The flow of time has no significance in the logically fixed pattern demanded by deterministic theory, time being a mere coordinate. In a theory with indeterminacy, however, the passage of time transforms the statistical expectations into real events.[25]

Roger Penrose, John Eccles and others have sought an explanation for the flow of time in the operation of the human brain itself, with the bold claim that some cerebral processes are irreducibly quantum-mechanical in nature. Though there is little hard experimental evidence at this stage to support such a theory, it nevertheless constitutes a fascinating avenue of research.

Attempts to explain the flow of time using physics, rather than trying to define it away using philosophy, are probably the most exciting contemporary developments in the study of time. Elucidating the mysterious flux would, more than anything else, help unravel the deepest of all scientific enigmas—the nature of the human self. Until we have a firm understanding of the flow of time, or incontrovertible evidence that it is indeed an illusion, then we will not know who we are, or what part we are playing in the great cosmic drama.

CHAPTER 14
THE UNFINISHED REVOLUTION

What seest thou else
In the dark backward and abysm of time?

WILLIAM SHAKESPEARE

You ain't seen nuthin' yet.

PRESIDENT RONALD REAGAN

David Deutsch once remarked that the history of science is the story of physics hijacking topics from philosophy. The nature of motion, the structure of the cosmos and the existence of atoms, for example, began as purely intellectual propositions discussed by Greek philosophers. Today they form part of mainstream physics. Even geometry, once thought to belong to a purely Platonic realm of idealized mathematical forms, became an experimental science with the general theory of relativity. The subject of consciousness may be the next on the list.

The nature of time was one of the central founding topics of early philosophical thought, and it dominated scholarly debate for centuries. The mysteries of time spread far beyond philosophy, however, into religion and politics, and eventually science, where for three hundred years time was treated simply as a conceptual "given," robbed of its subjective trappings. In 1905, Einstein plucked time from philosophy and placed it at the heart of physics. It suddenly became a physical thing, subject to laws and equations, and inviting experimental investigation. Nearly a century later, our understanding of time has advanced enormously, yet the Einstein revolution was clearly just the beginning. We are still a long way from solving the riddle of time.

So what are the major unanswered questions in the never-ending story of time? The following is my personal list of a dozen outstanding puzzles to be addressed (not in order of importance).

1. Tachyons: can we rule them out?

The special theory of relativity has been tested to unprecedented accuracy, and appears unassailable. Yet tachyons are a problem. Though they are allowed by the theory, they bring with them all sorts of unpalatable properties. Physicists would like to rule them out once and for all, but lack a convincing nonexistence proof. Until they construct one, we cannot be sure that a tachyon won't suddenly be discovered.

2. Black holes: do they really exist?

The most dramatic prediction of the general theory of relativity is undoubtedly the black hole, yet we await definitive confirmation that infinite timewarps exist in the real universe. Astronomers are searching hard, and the evidence for black holes continues to accumulate. I personally would be astonished if they do not exist. If they do, a host of questions follows. Is there really an end to time—a singularity—at the center of all black holes? Can black holes form tunnels or bridges to other universes, or even turn into wormholes that thread back into our universe? What happens to matter that falls into them? Are there such things as white holes?

3. Time travel: just a fantasy?

The investigation of exotic spacetimes that seem to permit travel into the past will remain an active field of research. So far, the loophole in the known laws of physics that permits time travel is very small indeed. Realistic time-travel scenarios are not known at the time of writing. But as with tachyons, in the absence of a no-go proof, the possibility has to stay on the agenda. So long as it does, paradoxes will haunt us.

4. Quantum questions.

The quantum domain is a wonderland of weird and perplexing temporal teasers. Time plays a very basic role in quantum physics, yet it enters the theory in a unique way that singles it out for special treatment—and special puzzlement. The relativity of time fits uneasily into the quantum picture of a world where transitions, and the "concretization" or "collapse" associated with measurements, apparently occur abruptly, at specific moments. Trouble comes when quantum states get entangled over extended spatial regions and simultaneous observations are made. The measurement of time itself is fraught with problems, because clocks are physical objects afflicted by quantum fuzziness.

5. Is time just a relic?

The difficulties are particularly acute when it comes to applying quantum mechanics to gravitation, for then the very spacetime continuum is subject to quantum fuzziness. The experts are divided about the need to pin down a sort of "master" time, a natural measure of change in a physically uncertain world, or to define time completely out of existence. The mystery of the vanishing time suggests to some that time is destined to be abandoned as a fundamental physical entity, a proposal that strikes others as outrageous and absurd. Could it be that, after millennia of deliberation about time, we shall finally discover that it doesn't really exist as a basic ingredient of reality, but is just some approximate property of a particular quantum state that happens to have been left over from the big bang?

6. The origin of time.

The fashionable theory that time originated with the big bang is probably the biggest outstanding issue, begging all sorts of (maybe unanswerable) questions concerning causality, God and eternity. If time existed before the big bang, we have to explain what physical processes predated this dramatic and violent event, and how it was caused. If the universe has always existed, we also run into major problems over the arrow of time. If, on the other hand, time really did "switch on" at the big bang, perhaps as a result of quantum processes, then we confront some equally tough problems. If the process was unique, can it be considered in any sense natural (as opposed to supernatural)? If it was not unique, and spacetimes can originate willy-nilly, are we forced to believe in an infinity of universes and an infinity of times?

7. The age of the universe.

The thorny problem of the age of the universe is right back at the head of the agenda. Taken at face value, the measurements of the expansion rate of the universe and the results of COBE, combined with realistic assumptions about dark matter, lead to the absurd conclusion that there are objects in space older than the universe. If so, the entire big-bang theory is suspect. With a bit of fudging and fitting, the problem has been swept under the carpet, because the observations are still fairly inaccurate. However, all that is about to change. With the Hubble telescope now fully operational, it should not be too long before we have a much better figure for the Hubble constant (the present expansion rate). If the value comes out above about 70, we're in real trouble. Keep watching the news reports!

8. The cosmological term: blunder or triumph?

Distasteful though many scientists find the cosmological term in Einstein's equations, there is no known reason to rule it out. If the forthcoming observations confirm the time-scale difficulty, then Einstein's greatest mistake will provide a spectacular and ready-made way of retaining the big-bang theory. If it is not needed for that purpose, this doesn't prove its nonexistence. The "cosmological-constant problem" (is it zero, and if so why?) will still need to be solved.

9. Beyond the standard theory?

Few physicists believe that Einstein's general theory of relativity is the last word on time. Quite apart from the problems of combining it with quantum mechanics, there are doubts that it will continue to apply all the way down to spacetime singularities, or in exotic circumstances where time loops threaten. Superaccurate testing of the theory using rapidly improving clock technology is probably the best way to probe its limits. In particular, the possibility that more than one time scale may exist must be tested. If there is a multiplicity of times, then the implications for cosmology and the age-of-the-universe problem are profound.

10. The arrow of time.

The mystery of time's arrow is the oldest problem in science concerning the nature of time, predating even the theory of relativity. It is intimately connected with the issue of the origin and possible end of the universe. Most scientists are agreed that the source of the asymmetry—i.e., time's directionality—can ultimately be traced to cosmology and the large-scale behavior of the universe, but the precise nature of the connection remains obscure and contentious. The theory that there may exist spacetime regions where time "runs backwards," or that the entire universe may be time-symmetric or even cyclic in time, is still popular in some quarters. There is plenty of scope for further investigation—and disagreement.

11. Time-symmetry violation.

The discovery that kaons break time-reversal symmetry has spawned many searches for T violation in other areas of particle physics, so far without success. The search for electric-dipole moments in the neutron and in various molecules promises to clarify the enigma of how past-future symmetry gets violated, and what relationship it may have to the arrow of time in cosmology.

12. The flow of time: mind or matter?

In my opinion, the greatest outstanding riddle concerns the glaring mismatch between physical time and subjective, or psychological, time. Experiments on human time perception are in their infancy; we have much to learn about the way the brain represents time, and how that relates to our sense of free will. The overwhelming impression of a flowing, moving time, perhaps acquired through a mental "back door," is a very deep mystery. Is it connected with quantum processes in the brain? Does it reflect an objectively real quantity of time "out there" in the world of material objects that we have simply overlooked? Or will the flow of time be proved to be entirely a mental construct—an illusion or a confusion—after all?

It is my personal belief that we are approaching a pivotal moment in history, when our knowledge of time is about to take another great leap forward. Einstein left us an important legacy. He showed us how time is part of the physical world, and gave us a magnificent theory that interweaves time with space and matter. Throughout the twentieth century, scientists have diligently explored the consequences of Einstein's time, both theoretically and experimentally. In doing so, they have unearthed some unnerving and bizarre possibilities, many of which have turned out to be true. Yet they have also encountered severe obstacles to a full understanding of time, hinting that Einstein's revolution remains unfinished. I believe its completion will prove a major outstanding challenge to twenty-first-century science.

EPILOGUE

... the greatest scientist of modern times.

THE TIMES OF LONDON, 21 APRIL 1955

Albert Einstein died on 18 April 1955. His health had been deteriorating for a decade, and an exploratory operation in 1948 had revealed a large aneurysm of the aorta, which caused him periods of acute abdominal pain. He spent his postwar years at the Institute of Advanced Study in Princeton, living and working in relative seclusion. Photographs of him reveal a slightly sad and weary look. Although he was a familiar figure in Princeton, he became increasingly distant from his fellow scientists. He showed little interest in the exciting discoveries in particle physics that characterized the early 1950s, and remained implacably opposed to quantum mechanics. His great preoccupation was the formulation of a unified field theory that would combine the various fundamental forces into a single mathematical scheme, and remove the appearance of indeterminacy from nature.

Einstein retained his interest in Zionism and world politics to the end. In 1952, he was formally offered the presidency of Israel by Ben-Gurion, but declined on grounds of a lack of suitable abilities. His lifelong pacifism and disgust at the construction of the atomic bomb made him a tireless campaigner for the abolition of nuclear weapons and for rapprochement with the Soviet Union.

Einstein was never very successful in his personal relationships. He remained estranged from his first wife and rarely saw his sons. Though he was closer to his second wife, Elsa, and her two daughters, when Elsa died in 1936, Einstein did not seem too upset. Indeed, he rather callously said he felt "more at home" with his mate gone. It gave him the opportunity to concentrate on his work, and he redoubled his efforts to find a unified field theory. His first wife, Mileva, died in 1948.

Following Elsa's death, Einstein's family in Princeton consisted of his sister, who died in 1951, his stepdaughter and his trusted assistant, Helen Dukas. When the aging scientist collapsed at home on 12 April 1955, it was Helen who attended to the crisis. The aortic aneurysm had ruptured and several days of hospitalization failed to stem a serious hemorrhage. Death became inevitable. For half a century, humanity's most influential scientist had dazzled the world with his scintillating intellect. Now that chapter of history was brought to a close. The man who had shown the world how time could be stretched had finally run out of it.

NOTES

Prologue

1. *Confessions* by St. Augustine, trans. R. S. Pine-Coffin (Penguin, Baltimore, 1961), 11:14, p. 294.

Chapter 1: A Very Brief History of Time

1. *The Roman Poet of Science Lucretius: De Rerum Natura*, set in English verse by A. D. Winspear (Harbor Press, New York, 1956), p. 22.
2. *The Book of Angelus Silesius*, trans. F. Franck (Vintage Books, New York, 1976), p. 45.
3. Ibid., p. 42.
4. *The Works of Plato* by B. Jowett (Oxford University Press, Oxford, 3rd ed. 1892), vol. 3, p. 456.
5. Augustine, *Confessions*, 11:13, p. 263.
6. *Scientific Theory and Religion* by E. W. Barnes (Cambridge University Press, Cambridge, 1933), p. 620.
7. *Foundations of Tibetan Mysticism* by Lama Anagarika Govinda (Samuel Weiser, New York, 1969), p. 116.
8. "Metaphysics of Time in Indian Philosophy and Its Relevance to Particle Science" by R. Reyna, in *Time in Science and Philosophy*, ed. J. Zeman (Academia, Prague, 1971), p. 238.
9. Ibid., pp. 233–34.
10. "The Dreaming" by W. E. H. Stanner, in *Traditional Aboriginal Society*, ed. W. H. Edwards (Macmillan, Melbourne, 1987), p. 225.
11. *Man and Time* by J. B. Priestley (Aldus Books, London, 1964), p. 141.
12. *The Myth of the Eternal Return* by M. Eliade, trans. W. R. Trask (Pantheon Books, New York, 1954), p. ix.
13. Ibid., p. 34.
14. "Evolution, Myth and Poetic Vision" by W. J. Ong, in *The Enigma of Time*, ed. P. T. Landsberg (Adam Hilger, Bristol, Eng., 1982), p. 220.
15. *John Harrison: The Man Who Found Longitude* by H. Quill (John Baker Publishers, London, 1966), p. 6.

16. *The Mathematical Principles of Natural Philosophy* by I. Newton, trans. A. Motte (University of California Press, Berkeley, 1962), vol. 1, p. 7.

17. "The Rediscovery of Time" by I. Prigogine, in *Science and Complexity*, ed. S. Nash (Science Reviews, Northwood, Middlesex, 1985), p. 11.

18. *On the Origin of Species by Means of Natural Selection* by C. Darwin (John Murray, London, 2nd ed. 1860), p. 486.

19. *Eternal Recurrence* by F. Nietzsche, in *The Complete Works of Friedrich Nietzsche*, ed. O. Levy (G. T. Foulis, Edinburgh, 1910).

20. Philo in *Quod Deus Immutabilis Sit* 6:32, ed. L. Cohn and P. Wendland (Macmillan, London, 1896), vol. 2, p. 63.

21. *October the First Is Too Late* by F. Hoyle (Heinemann, London, 1966), pp. 75–82.

Chapter 2: Time for a Change

1. "On the Electrodynamics of Moving Bodies" by A. Einstein, reprinted in English in *Einstein: A Centenary Volume*, ed. A. P. French (Harvard University Press, Cambridge, Mass., 1979), p. 281.

2. Quoted in *Reality and Scientific Truth* by I. Rosenthal-Schneider (Wayne State University Press, Detroit, 1980), p. 74.

3. Quoted in *Subtle Is the Lord: The Science and the Life of Albert Einstein* by A. Pais (Oxford University Press, Oxford, 1982), p. 139.

4. *Science at the Crossroads* by H. Dingle (Martin Brian & O'Keefe, London, 1972), p. 143.

5. Ibid., p. 17.

6. Ibid.

7. Ibid.

8. Quoted in *Time and the Novel* by A. A. Mendilow (Peter Nevill, New York, 1952), p. 72.

9. *Parerga and Paralipomena: Short Philosophical Essays* by A. Schopenhauer, trans. E. F. J. Payne (Clarendon Press, Oxford, 1974), p. 283.

10. *Confessions* by St. Augustine, trans. R. S. Pine-Coffin (Penguin, Baltimore, 1961), p. 253.

11. Quoted in *Relativity for Scientists and Engineers* by R. Skinner (Dover, New York, 1982), p. 27.

12. *Jerusalem* by W. Blake, 15:6.

13. "Burnt Norton," in *Collected Poems 1909–1962* by T. S. Eliot (Faber & Faber, London, 4th ed. 1963), p. 194.

14. Quoted in Pais, *Subtle Is the Lord*, p. 152.

15. "What Is the Fourth Dimension?" in *Scientific Romances* by C. H. Hinton (Swan Sonnenschein, London, 1884), p. 34.

16. Ibid.

17. *Philosophy of Mathematics and Natural Science* by H. Weyl (Princeton University Press, Princeton, 1949), p. 122.

18. Eliot, "Burnt Norton," p. 189.

19. Weyl, *Philosophy of Mathematics*, p. 166.

20. Quoted in *The Philosophy of Rudolf Carnap*, ed. P. A. Schilpp (Open Court, La Salle, Ill., 1963), p. 37.

Chapter 3: Timewarps

1. "On the Electrodynamics of Moving Bodies" by A. Einstein, reprinted in English in *Einstein: A Centenary Volume*, ed. A. P. French (Harvard University Press, Cambridge, Mass., 1979), p. 292.
2. *The Perpetual Motion Mystery: A Continuing Quest* by R. A. Ford (Lindsay Publications, Bradley, Ill., 1987), p. 41.
3. Quoted in *The Laboratory of the Mind* by J. R. Brown (Routledge, London, 1991), ch. 5.
4. Quoted in *Subtle Is the Lord: The Science and the Life of Albert Einstein* by A. Pais (Oxford University Press, Oxford, 1982), p. 448.

Chapter 4: Black Holes: Gateways to the End of Time

1. *Philosophical Transactions of the Royal Society (London)*, vol. 74 (1784), p. 35.
2. *Monthly Notices of the Royal Astronomical Society*, vol. 80 (1920), p. 96.
3. Quoted in *The Portable Curmudgeon* by J. Winokur (NAL Books, New York, 1987), p. 157.
4. "Dark Stars: The Evolution of an Idea" by W. Israel, in *300 Years of Gravitation*, ed. S. W. Hawking and W. Israel (Cambridge University Press, Cambridge, 1987), p. 206.
5. "On a Stationary System with Spherical Symmetry Consisting of Many Gravitating Masses," *Annals of Mathematics*, by A. Einstein, vol. 40 (1939), p. 922.
6. *Space, Time and Gravitation* by A. S. Eddington (Cambridge University Press, Cambridge, 1920), p. 98.
7. *Physical Review*, vol. 56 (1939), p. 455.
8. Israel, "Dark Stars," p. 231.
9. *Statistical Physics* by L. Landau and E. M. Lifshitz, trans. E. and R. F. Peierls (Pergamon, London, 1958), p. 343.
10. *Philosophical Magazine*, vol. 39 (1920), p. 626.
11. *Black Holes and Time Warps* by K. S. Thorne (Norton, New York, 1994), p. 255.
12. Ibid., p. 239.

Chapter 5: The Beginning of Time: When Exactly Was It?

1. "On the Beginning of Time," in *The City of God* by St. Augustine of Hippo, trans. H. Bettenson (Penguin, Harmondsworth, 1972).
2. *Cosmology* by H. Bondi (Cambridge University Press, Cambridge, 1952), p. 165.
3. Quoted in *Einstein: A Life in Science* by Michael White and John Gribbin (Simon & Schuster, London, 1993), p. 203.
4. "Personal Recollections: Some Lessons for the Future" by W. McCrea, in *Modern Cosmology in Retrospect*, eds. R. Bertotti, R. Balbinot, S. Bergia and A. Messina (Cambridge University Press, Cambridge, 1990), p. 207.

Chapter 6: Einstein's Greatest Triumph?

1. See "Dark Matter" by J. Trefil in *Smithsonian* (June 1993), p. 27.
2. "The Extragalactic Universe: an Alternate View," by H. C. Arp, G. Bur-

bidge, F. Hoyle and N. C. Wickramasinghe, in *Nature*, vol. 346 (1990), p. 810.
3. "The Cosmological Constant" by S. W. Hawking, *Philosophical Transactions of the Royal Society (London) A*, vol. 310 (1983), p. 303.
4. *Dreams of a Final Theory* by S. Weinberg (Random House, New York, 1992), p. 224.

Chapter 7: Quantum Time

1. "Quantum Optical Tests of Complementarity" by M. O. Scully, B. G. Englert and H. Walter, in *Nature*, vol. 351 (1991), p. 111.
2. "Observation of a 'Quantum Eraser': a Revival of Coherence in a Two-Photon Interference Experiment" by P. G. Kwait, A. M. Steinberg and R. Y. Chiao, in *Physical Review A*, vol. 45 (1992), p. 7729.
3. "Induced Coherence and Indistinguishability in Optical Interference" by X. Y. Zhou, L. J. Wang and L. Mandel, in *Physical Review Letters*, vol. 67 (1991), p. 318.
4. "Faster Than Light?" by R. Y. Chiao, P. G. Kwait and A. M. Steinberg, *Scientific American* (August 1993), p. 38.
5. "God, Time and the Creation of the Universe" by C. Isham, in *Explorations in Science and Theology*, ed. E. Winder (RSA, London, 1993), p. 58.

Chapter 8: Imaginary Time

1. *Correspondence* of Leibniz–Clarke, Leibniz's 4th paper, sect. 15.
2. For a discussion of Kant's antinomies, see *The Measure of the Universe* by J. D. North (Clarendon Press, Oxford, 1965), pp. 390–91.
3. *The Physics of Immortality* by F. Tipler (Doubleday, New York, 1994).
4. "Time Without End: Physics and Biology in an Open Universe" by F. Dyson, *Reviews of Modern Physics*, vol. 51 (1979), p. 447.

Chapter 9: The Arrow of Time

1. W. Ritz and A. Einstein, in *Physikalische Zeitschrift*, vol. 10 (1909), p. 323.
2. "Interaction with the Absorber as the Mechanism of Radiation" by J. A. Wheeler and R. P. Feynman, *Reviews of Modern Physics*, vol. 17 (1945), p. 157.
3. "Absorber Theory of Radiation and the Future of the Universe" by R. B. Partridge, *Nature*, vol. 244 (1973), p. 263.
4. "Causality and Faster Than Light Particles" by P. L. Csonka, *Nuclear Physics B*, vol. 21 (1970), p. 436.
5. *Time Machines* by P. J. Nahin (American Institute of Physics, New York, 1993), p. 225.
6. "Can Time Go Backward?" by M. Gardner, *Scientific American*, vol. 216, no. 1 (1967), p. 6.
7. *Almanach des Lettres Françaises et Étrangères* by H. Berlioz, reprinted in *Larousse des Citations Françaises et Étrangères* (Larousse, Paris, 1976), p. 68.
8. "CP and CPT Symmetry Violations, Entropy and the Expanding Universe" by Y. Ne'eman, in *International Journal of Theoretical Physics*, vol. 3 (1970), p. 1.

Chapter 10: Backwards in Time

1. *The Sophist and The Statesman* by Plato, ed. A. E. Taylor (Nelson, London, 1961), p. 277.
2. "The Arrow of Time" by T. Gold in *Time*, eds. S. T. Butler and H. Messel (Shakespeare Head Press Proprietary, Sydney, 1965), p. 159.
3. Ibid., p. 161.
4. "Can Time Go Backward?" by M. Gardner, *Scientific American*, vol. 216, no. 1 (1967), p. 2.
5. *Cybernetics* by N. Wiener (MIT Press, Cambridge, Mass., 1948), p. 45.
6. "Time Symmetric Electrodynamics and the Arrow of Time" by F. Hoyle and J. V. Narlikar, *Proceedings of the Royal Society (London) A*, vol. 277 (1964), p. 1.
7. "The No-Boundary Condition and the Arrow of Time" by S. Hawking, in *The Physical Origins of Time Asymmetry*, ed. J. J. Halliwell, J. Perez-Mercader and W. H. Zurek (Cambridge University Press, Cambridge, 1994), p. 346.
8. "Cosmology, Time's Arrow, and That Old Double Standard" by H. Price, in *Time's Arrows Today*, ed. S. Savitt (Cambridge University Press, Cambridge, 1994).

Chapter 11: Time Travel: Fact or Fantasy?

1. *The Nature of the Physical World* by A. S. Eddington (Cambridge University Press, Cambridge, 1929) pp. 57–58.
2. "An Example of a New Type of Cosmological Solution of Einstein's Field Equations of Gravitation" by K. Gödel, in *Reviews of Modern Physics*, vol. 21 (1949), p. 447.
3. Cited in ibid.
4. "Gravitational Field of a Spinning Mass as an Example of Algebraically Special Metrics" by R. Kerr, in *Physical Review Letters*, vol. 11 (1963), p. 237.
5. "Rotating Cylinders and the Possibility of Global Causality Violation" by F. J. Tipler, in *Physical Review D*, vol. 9 (1974), p. 2203.
6. *Contact* by C. Sagan (Simon & Schuster, New York, 1985).
7. *Black Holes and Time Warps* by K. S. Thorne (Norton, New York, 1994).
8. "Closed Timelike Curves Produced by Pairs of Moving Cosmic Strings: Exact Solutions" by J. R. Gott III, in *Physical Review Letters*, vol. 66 (1991), p. 1126.
9. "Quantum Mechanics and Closed Timelike Lines" by D. Deutsch, in *Physical Review D*, vol. 44 (1991), p. 3197.
10. "Chronology Protection Conjecture" by S. W. Hawking, in ibid., vol. 46 (1992), p. 603.
11. *The Time Machine* by H. G. Wells (Heinemann, London, 1895), p. 151.

Chapter 12: But What Time Is It *Now?*

1. "Time and Becoming" by J. J. C. Smart, in *Time and Cause*, ed. P. van Inwagen (Reidel, Dordrecht, 1980), pp. 3–15.
2. "The Unreality of Time" by J. E. McTaggart in *Mind*, vol. 17 (1908), p. 457.

3. *An Experiment with Time* by J. W. Dunne (Faber & Faber, London, 1927).
4. "The Myth of the Passage of Time" by D. Park, *Studium Generale*, vol. 24 (1971), p. 20.
5. "Dirac's Cosmology and Mach's Principle" by R. H. Dicke, in *Nature*, vol. 192 (1961), p. 440.
6. "The Anthropic Principle and Its Implications for Biological Evolution" by B. Carter, *Philosophical Transactions of the Royal Society of London A*, vol. 310 (1983), p. 347.
7. "Time and the Anthropic Principle" by J. Leslie, *Mind*, vol. 101 (1992), p. 403.

Chapter 13: Experimenting with Time

1. *The Dimension of the Present Moment and Other Essays* by M. Holub, ed. David Young (Faber & Faber, London, 1990), p. 6.
2. For a full description, see, for example, *Consciousness Explained* by D. C. Dennett (Little, Brown, London, 1991), chh. 5, 6.
3. *Ways of Worldmaking* by N. Goodman (Harvester, Sussex, 1983), pp. 73–74.
4. Dennett, *Consciousness Explained*, p. 115.
5. Ibid., p 127.
6. "Subjective Referral of the Timing for a Conscious Sensory Experience" by B. Libet, E. W. Wright, Jr., B. Feinstein and D. K. Pearl, *Brain*, vol. 102 (1979), p. 193.
7. "Voluntary Finger Movements in Man: Cerebral Potentials and Theory" by L. Deeke, B. Grotzinger and H. H. Kornhuber, *Biological Cybernetics*, vol. 23 (1976) p. 99.
8. *The Self and Its Brain* by K. Popper and J. Eccles (Springer International, New York, 1977).
9. *The Emperor's New Mind* by R. Penrose (Oxford University Press, Oxford, 1989).
10. Dennett, *Consciousness Explained*, chh. 5, 6.
11. *The Panda's Thumb* by S. J. Gould (Norton, New York, 1992), p. 251.
12. *Dragon's Egg* by R. Foreword (Ballantine, New York, 1980).
13. "Subjective Time" by S. Albert in *The Study of Time III*, eds. J. T. Fraser, N. Lawrence and D. Park (Springer-Verlag, New York, 1978), p. 269.
14. "Time and the Structure of Human Cognition" by M. Toda, in *The Study of Time II*, eds J. T. Fraser and N. Lawrence (Springer-Verlag, Berlin, 1975), p. 314.
15. *From Being to Becoming* by I. Prigogine (Freeman, San Francisco, 1980).
16. Penrose, *The Emperor's New Mind*.
17. *The Nature of the Physical World* by A. S. Eddington (Cambridge University Press, Cambridge, 1929), p. 97.
18. Ibid., p. 51.
19. Penrose, *The Emperor's New Mind*, p. 304.
24. *A Philosophical Essay on Probabilities* by P. S. Laplace (Dover, New York, 1951), p. 4 (original publication, 1819).
25. "Relativity and Indeterminacy" by H. Bondi, *Nature*, vol. 169 (1952), p. 660.

BIBLIOGRAPHY

Aveni, A. *Empires of Time*. Basic Books, New York, 1989.

Barrow, J., and Tipler, F. *The Anthropic Cosmological Principle*. Oxford University Press, Oxford, 1986.

Capek, M., ed. *The Concepts of Space and Time*. Reidel, Dordrecht, 1976.

Clark, R. *Einstein: The Life and Times*. Hodder & Stoughton, London, 1973.

Coveney, P., and Highfield, R. *The Arrow of Time*. W. H. Allen, London, 1990.

Davies, P. C. W. *The Physics of Time Asymmetry*. University of California Press, Berkeley and Los Angeles, 1974.

——. *The Mind of God*. Simon & Schuster, New York, 1991.

——. *The Edge of Infinity*. Rev. ed., Penguin, London, 1994.

——. *The Last Three Minutes*. Basic Books, New York, 1994.

Eliade, M. *The Myth of the Eternal Return*. Routledge & Kegan Paul, London, 1955.

Flood, R., and Lockwood, M. *The Nature of Time*. Basil Blackwell, Oxford, 1986.

Fraser, J. T. *The Genesis and Evolution of Time*. University of Massachusetts Press, Amherst, 1982.

——. *Time: The Familiar Stranger*. Tempus, Washington, D.C., 1987.

——, ed. *The Voices of Time*. Brazilier, New York, 1966.

French, A. P. *Einstein: A Centenary Volume*. Harvard University Press, Cambridge, Mass., 1979.

Gale, R. M., ed. *The Philosophy of Time*. Macmillan, London, 1968.

Gold, T., ed. *The Nature of Time*. Cornell University Press, Ithaca, N.Y., 1967.

Gould, S. J., *Time's Arrow, Time's Cycle*. Penguin, Harmondsworth, 1988.

Gribbin, J. *Timewarps*. Delacorte, New York, 1979.

Grunbaum, A. *Philosophical Problems in Space and Time*. Knopf, New York, 1963.

Halliwell, J. J., Perez-Mercader, J., and Zurek, W. H., eds. *The Physical Origins of Time Asymmetry*, Cambridge University Press, Cambridge, 1994.

Hawking, S. W. *A Brief History of Time*. Bantam, New York, 1988.

Highfield, R., and Carter, P. *The Private Lives of Albert Einstein*. Faber & Faber, London, 1993.

Landsberg, P., ed. *The Enigma of Time*. Adam Hilger, Bristol, Eng., 1982.

Leslie, J. "Testing the Doomsday Argument." *Journal of Applied Philosophy*, vol. 11 (1994), p. 31.

Luminet, J. P. *Black Holes*. Cambridge University Press, Cambridge, 1992.

Morris, R. *Time's Arrows*. Simon & Schuster, New York, 1984.

Mowaljarlai, D., and Malnic, J. *Yorro Yorro: Spirit of the Kimberley*. Magabala Books, Broome, Australia, 1993.

Nahin, P. J. *Time Machines*. American Institute of Physics, New York, 1993.

North, J. D. *The Measure of the Universe*. Clarendon Press, Oxford, 1965.

Pais, A. *Subtle Is the Lord: The Science and the Life of Albert Einstein*. Oxford University Press, Oxford, 1982.

Penrose, R. *The Emperor's New Mind*. Oxford University Press, Oxford, 1989.

Pike, N. *God and Timelessness*. Routledge & Kegan Paul, London, 1970.

Priestley, J. B. *Man and Time*. Aldus Books, London, 1964.

Prigogine, I. *From Being to Becoming*. Freeman, San Francisco, 1980.

Pyenson, L. *The Young Einstein*. Adam Hilger, Bristol, Eng., 1985.

Reichenbach, H. *The Direction of Time*. University of California Press, Berkeley, 1956.

Sachs, R. *The Physics of Time Reversal*. University of Chicago Press, Chicago, 1987.

Shallis, M. *On Time*. Schocken, New York, 1983.

Smart, J. J. C., ed. *Problems of Space and Time*. Macmillan, New York, 1964.

Thorne, K. S. *Black Holes and Time Warps*. Norton, New York, 1994.

Tolman, R. C. *Relativity, Thermodynamics and Cosmology*. Clarendon Press, Oxford, 1934.

Wald, R. M. *Space, Time and Gravity*. University of Chicago Press, Chicago, 1977.

Wheeler, J. A. *Frontiers of Time*. North Holland, Amsterdam, 1979.

White, M., and Gribbin, J. *Einstein: A Life in Science*. Simon & Schuster, London, New York, 1993.

Whitrow, G. J. *What Is Time?* Thames and Hudson, London, 1972.

———. *The Natural Philosophy of Time*. Oxford University Press, Oxford, 2nd ed. 1980.

———. *Time in History*. Oxford University Press, Oxford, 1989.

Will, C. *Was Einstein Right?* Basic Books, New York, 1986.

Winfree, A. *The Timing of Biological Clocks*. Freeman, San Francisco, 1987.

Yates, J. C. *The Timelessness of God*. University Press of America, Lanham, Md., 1990.

Zeh, H. D. *The Physical Basis of the Direction of Time*. Springer-Verlag, Berlin, 1989.

INDEX

Page numbers in *italics* refer to captions.

Abano, Petro d', 126
aborigines, Australian, 23, 26–27
absolute zero, 97
acceleration:
  in general theory of relativity,
    102
  gravity vs., 87–88, 90, 93
  uniform motion vs., 48, 52, 59
action at a distance, 175, 176
Adams, Walter, 109
Addison, Joseph, 23
Adelaide, University of, 163, 252
Advaita Vedanta, 25–26
advanced waves:
  as going back in time, 202–3, 209
  interference and, 202
  probability of, 200, 201, 209
  retarded waves vs., 197
  symmetry and, 200, 209
Agathon, 168
age of the universe. *See* universe,
  age of
aging process:
  time dilation and, 59–60, 82–83,
    115
  time reversal and, 200
Albert, Stuart, 273–74
Amis, Martin, 29, 255
Anderson, A., 116–17
Andromeda (M31) nebula, 129, 130,
  131

Angelus Silesius, 23
anisotropy, dipole, 128–29
Annalen der Physik, 47
Ann and Betty, *see* twins effect
anthropic principle, 259
antigravity:
  quantum, 247–48
  see also cosmological constant
antimatter, 204–18
  first prediction and discovery of,
    204–5
  time reversal and, 204–7, *205,*
    207, 213–18, *215*
  see also kaons
antiquarks, 208, 211
antiworlds, 224–26, 228, 232, 244
Arecibo radio telescope, 45
Aristotle, 29–30, 94
Arp, Halton, 149, 150–51, 156
arrow of time, 196–232
  antimatter and, 204–18, *206, 211*
  antiworlds and, 224–26, 228, 232
  biological evolution and, 35, 36
  contracting universe and, 219–32,
    221, 228, 229
  cosmic-scale reversals of, 227–29,
    228, 229
  definition of, 15, 196
  electromagnetic waves and, 196–
    204, 209
  gravity and, 217–18